D1243238

Discrete Models

Applied Mathematics and Computation

A Series of Graduate Textbooks, Monographs, Reference Works

Series Editor: ROBERT KALABA, University of Southern California

No. 1 MELVIN R. SCOTT
Invariant Imbedding and its Applications to Ordinary Differential Equations: An Introduction, 1973

No. 2 JOHN CASTI and ROBERT KALABA
Imbedding Methods in Applied Mathematics, 1973

No. 3 DONALD GREENSPAN
Discrete Models, 1973

Other volumes in preparation

Discrete Models

DONALD GREENSPAN
Department of Computer Sciences
and
Madison Academic Computing Center
University of Wisconsin
Madison, Wisconsin

1973
Addison-Wesley Publishing Company
Advanced Book Program
Reading, Massachusetts

London · Amsterdam · Don Mills, Ontario · Sydney · Tokyo

CODEN: APMCC

Library of Congress Cataloging in Publication Data

Greenspan, Donald. 1928-
Discrete models.

(Applied mathematics and computation, no. 3)
Bibliography: p.
1. Electronic data processing--~~Difference equations.~~ ~~2. Electronic data processing--Nonlinear theories.~~ ~~3. Electronic data processing--Mathematical physics.~~ I. Title.
QA431.G75 515'.625 73-7957
ISBN 0-201-02612-0
ISBN0-201-02613-9 (pbk.)

Reproduced by Addison-Wesley Publishing Company, Inc., Advanced Book Program, Reading, Massachusetts, from camera-ready copy prepared by the author.

American Mathematical Society (MOS) Subject Classification Scheme (1970): 39A10, 65L05, 68-02, 68A55, 70D05, 70E15, 70F05, 70F10, 70F15, 70K10, 70K15, 76F05, 76L05, 80A20, 83A05

Manufactured in the United States of America

ISBN 0-201-02612-0 (hardbound)
ISBN 0-201-02613-9 (paperback)
ABCDEFGHIJ-MA-79876543

"The discussion of the discrete approach to physical problems illustrates another aspect of the application of computers that is potentially more important than their use as glorified slide rules. This aspect is the following one: New models of physical (and other) phenomena may be created that are more adapted to computer analysis than are those of current physical theories."

A. H. Taub

Studies in Applied Mathematics
Mathematical Association of America
1971

CONTENTS

SERIES EDITOR'S FOREWORD

Execution times of modern digital computers are measured in nanoseconds. They can solve hundreds of simultaneous ordinary differential equations with speed and accuracy. But what does this immense capability imply with regard to solving the scientific, engineering, economic, and social problems confronting mankind? Clearly, much effort has to be expended in finding answers to that question.

In some fields, it is not yet possible to write mathematical equations which accurately describe processes of interest. Here, the computer may be used simply to simulate a process and, perhaps, to observe the efficacy of different control processes. In others, a mathematical description may be available, but the equations are frequently difficult to solve numerically. In such cases, the difficulties may be faced squarely and possibly overcome; alternatively, formulations may be sought which are more compatible with the inherent capabilities of computers. Mathematics itself nourishes and is nourished by such developments.

Each order of magnitude increase in speed and memory size of computers requires a reexamination of computational techniques and an assessment of the new problems which may be brought within the realm of solution. Volumes in this series will provide indications of current thinking regarding problem formulations, mathematical analysis, as well as computational treatment.

ROBERT KALABA

Los Angeles, California
April, 1973

PREFACE

This monograph is concerned primarily with a new, rich source of mathematical models. In it is developed a deterministic, yet discrete, approach to the approximation of phenomena involving large numbers of particles, each of which is still that significant a portion of matter that the classical laws of Newtonian mechanics are applicable. That such an approach is feasible is a direct consequence of the availability of modern, high-speed, digital computers, and of the observation that "...conditions in which the classical laws of momentum and energy fail perceptibly for tangible portions of matter are extremely rare if not altogether unknown."*

For permission to quote freely from various articles published previously, I wish to thank the publishers of the Computer Journal, the Journal of Mathematical Analysis and Applications, BIT, the Journal of the Franklin Institute, Utilitas Mathematica, the American Mathematical Monthly, the Journal of Computers and Structures, and Foundations of Physics.

Finally, since this book is being published by a photo-offset process from an original manuscript, credit for the typing should be given to Patricia Hanson and for the illustrations to Martha Fritz.

Donald Greenspan

Madison, Wisconsin

*C. Truesdell and R. A. Toupin
Encyclopedia of Physics
Springer Verlag
1960

Discrete Models

CHAPTER I - FUNDAMENTALS OF DISCRETE MODEL THEORY

1.1 INTRODUCTION

To deny the concept of *infinity* is as unmathematical as it is un-American. Yet, it is precisely a form of such mathematical heresy upon which discrete model theory is built.

There are a variety of compelling reasons why this basic concept of both classical and modern mathematical analysis should be considered suspect. *Geometrically*, the neophyte may be amused by the surface generated by rotating $y = \frac{1}{x}$, $1 \le x < \infty$, around the X-axis, since the resulting "infinite horn" has a finite volume, but an infinite surface area. It is often said, and, indeed, rather loosely, that one can fill the inside of this surface with a finite amount of paint, yet this quantity of paint is not sufficient to paint the outside. But much more disturbing examples than this exist, for it is possible to describe three-dimensional solids whose volumes are arbitrarily large, but whose surface areas are arbitrarily small (Besicovitch).

Physically, the concept of infinity has no known counterpart, since, for example, all solids and particles are believed to be finite from both the quantitative and qualitative points of view. It is unfortunate that so many scientists have been conditioned to believe that, say, 10^{30} particles can <u>always</u> be approximated well by an infinite number of points. For, indeed, to approximate a physical particle by a mathematical point is to neglect the rich structures of the atom and the molecule, while to approximate 10^{30} objects of any type by an infinite number of such objects is to have enlarged the given set by so great an amount that the given objects are <u>entirely negligible</u> in the enlarged set, or, more precisely, in any nondegenerate interval of real numbers, 10^{30} points form a set of measure zero (Halmos). Also, from the physical point of view, as models become more and more sophisticated, the study of dynamical behavior by means of continuous mathematical models invariably requires the solution of nonlinear differential equations, and despite the unbelievable volume of both classical and modern mathematics which exists, no general methods have ever emerged for solving such equations. Finally, from the physical point of view, it is worth noting that the concept of infinity has led to paradoxical results like those related to motion and described by Zeno as early as 300 B.C. These paradoxes have never been resolved in the sense that the fundamental problems have merely been shifted to more subtle, less obvious problems related to the convergence of <u>infinite</u> series (Whitehead).

From the modern <u>computer</u> point of view, the concept of infinity is completely foreign. Digital computers have largest numbers, smallest numbers, a finite number of numbers, and only numbers with a finite number of decimal places. The output from these computers is finite in every sense. But, further, if one examines the application of computers to approximating solutions of nonlinear differential equations by numerical methods, which is one of the most successful areas of computer applications, it is the concept of infinity, manifested in continuity, which demands the usually impossible task of proving that a numerical solution converges to an analytical solution as a grid size converges to zero (see, e.g., Greenspan (6), Henrici, Moore, Richtmyer and Morton, and Urabe).

From a philosophical point of view, the importance of nonlinear behavior and the application of digital computers to the study of such behavior have also created reasons for denying the concept of infinity. As shown in Figure 1.1, it is usual, first, in the development of scientific knowledge,

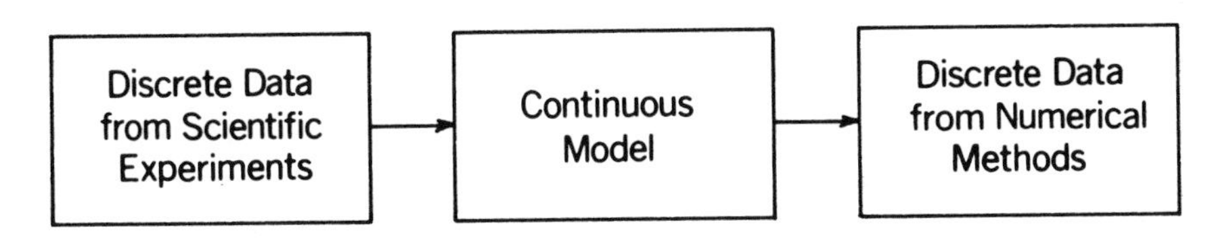

FIGURE 1.1

to have experimentation, which results in discrete sets of data. Theoreticians then analyze these data and, in the classical spirit, infer continuous models. Should the equations of these models be nonlinear, these would be solved today on computers by numerical methods, which results again in discrete data. Philosophically, the middle step of the activity sequence in Figure 1.1 is inconsistent with the other two steps. Indeed, it would be simpler and more consistent to replace the continuous model inference by a discrete model inference, as shown in Figure 1.2, and this can be accomplished by denying the concept of infinity.

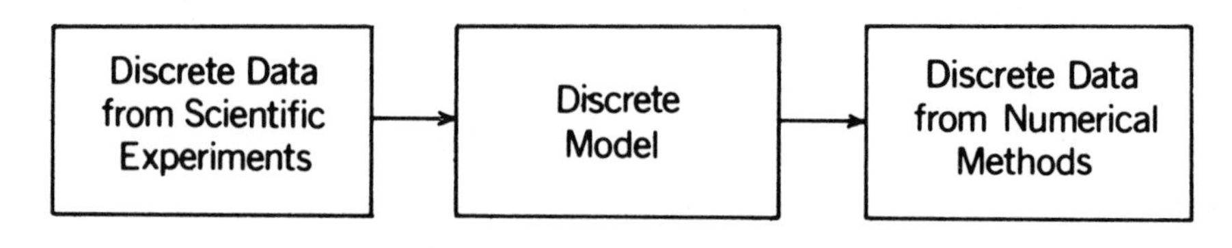

FIGURE 1.2

Motivated by the reasons listed above, we will proceed under the following mathematical and physical assumptions. The concept of infinity and the consequential concepts of limit, derivative, and integral are reasonable for the pure mathematical study of real numbers and real functions, but are not reasonable for the modeling of physical concepts and phenomena.

Our primary aim will be the study of nonlinear physical behavior. Classical continuous mathematics will be used only in the study of stability, where properties of sets of rational numbers can be derived most easily by considering these numbers as subsets of the real number system. Dynamical behavior will be studied entirely in terms of arithmetic, or, more precisely, in terms of high-speed arithmetic, for it is the availability of the modern digital computer which will make our approach both reasonable and practical. The dynamical equations of our models will be difference equations which, whether linear or nonlinear, will easily be solvable. Thereby, it is hoped that if an applied scientist is willing to learn the simple language of a computer, then he need be equipped only with the rudimentary mathematical knowledge of arithmetic and algebra in order to study highly complex physical phenomena.

1.2 BASIC MATHEMATICAL CONCEPTS

In this section, we will concentrate on basic concepts and theorems in one dimension only, for the extension to an arbitrary number of dimensions follows easily by the usual introduction of vectors.

<u>Definition 1.1.</u> For $\Delta x > 0$ and for a fixed constant a, the finite set $x_k = a + k\Delta x$, $k = 0,1,2,\ldots,n$, is called an R_{n+1} set.

<u>Example 1.</u> For $\Delta x = 0.25$, $a = 0$, and $n = 4$, the set $x_0 = 0$, $x_1 = 0.25$, $x_2 = 0.50$, $x_3 = 0.75$, $x_4 = 1$ is an R_5 set. It is the set of division points which results in classical mathematics by dividing the interval $0 \le x \le 1$ into four equal parts.

<u>Example 2.</u> For $\Delta x = 10^{-6}$, $a = 0$, and $n = 10^6$, the set $x_k = k(10)^{-6}$, $k = 0,1,2,\ldots,10^6$, is an R_{10^6+1} set. Without the aid of special lenses, the plot of this set appears to the naked eye to be no different than the classical real number interval $0 \le x \le 1$. This "packing" of a large, but finite, number of points to give the same visual effect as continuity, coupled with the ability to manipulate constructively with such sets by means of computers, makes R_{n+1} sets both mathematically and physically attractive.

<u>Definition 1.2.</u> Let $x_k = a + k\Delta x$, $k = 0,1,\ldots,n$, be an R_{n+1} set. If to each $x_i \in R_{n+1}$ there corresponds by some rule f a unique value y_i, then f is said to be a <u>discrete</u> function on R_{n+1}.

<u>Example 1.</u> On the R_{10^6+1} set $x_k = k(10)^{-6}$, $k = 0,1,2,\ldots,10^6$, the relationship $y_i = x_i^2$, $i = 0,1,2,\ldots,10^6$, defines a discrete function.

Example 2. On the R_5 set $x_0 = 0$, $x_1 = 0.25$, $x_2 = 0.50$, $x_3 = 0.75$, $x_4 = 1.00$, a discrete function is defined by the listing $y_0 = 1$, $y_1 = -1$, $y_2 = 0$, $y_3 = 7$, $y_4 = 3$.

Discrete functions can be defined by explicit relationships of the form $y_i = f(x_i)$, as given in Example 1, above, or by a complete listing, as given in Example 2, above. Utilization of a large computer memory bank for storage enables one to use experimental data directly without having to infer any explicit form of the dependence of y_i on x_i for each value of i.

Since our major aim will be the development of discrete concepts and phenomena, the term "function" will mean "discrete function" throughout, unless indicated otherwise.

Definition 1.3. Let f be a function defined on a given R_{n+1} set. Then the graph of f is the set of number couples (x_i, y_i), $i = 0,1,2,\ldots,n$.

Example. On the R_{10^6+1} set $x_k = k(10)^{-6}$, $k = 0,1,2,\ldots,10^6$, the graph of $y_i = x_i^2$ is given in Figure 1.3. Again, note that the process of packing a relatively large, but finite, number of points has resulted in a graph which, to the naked eye, does not differ from that of $y = x^2$ on $0 \leq x \leq 1$.

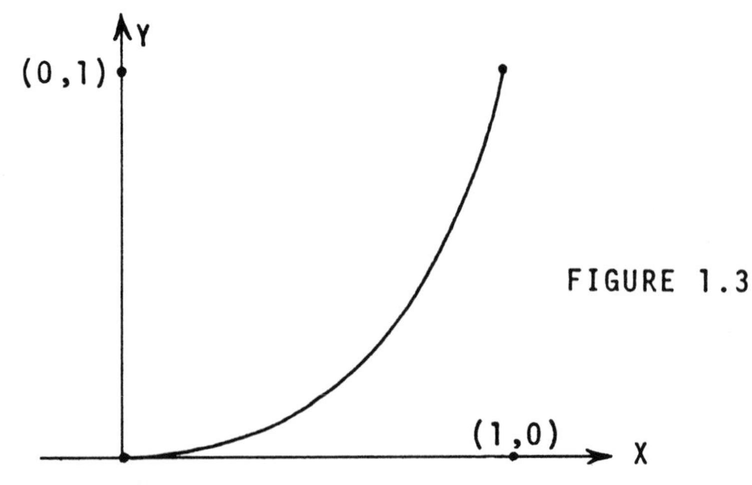

FIGURE 1.3

With the basic set and function concepts now defined, we turn to the concept of a first order difference equation. Intuitively, we wish to introduce an equation of the form

$$(1.1)\qquad \frac{y_{i+1}-y_i}{\Delta x} = f(x_i,y_i,y_{i+1}), \quad i = 0,1,2,\dots,n-1.$$

But the form of (1.1) is not very convenient from the computer point of view. However, if it is rewritten as

$$y_{i+1} = y_i + (\Delta x)f(x_i,y_i,y_{i+1}),\ i = 0,1,2,\dots,n-1,$$

or, equivalently, as

$$(1.2)\qquad y_{i+1} = F(x_i,y_i,y_{i+1}), \quad i = 0,1,2,\dots,n-1,$$

where

$$F(x_i,y_i,y_{i+1}) \equiv y_i + (\Delta x)f(x_i,y_i,y_{i+1}) ,$$

then the form (1.2) is exceptionally convenient, for (1.2) is a recursion formula and such formulas are especially suitable for high-speed computation. Our definition is then formulated as follows.

Definition 1.4. On an R_{n+1} set $x_0,x_1,\dots,x_n$, an equation of the form (1.2), where F must depend on y_i, but need not depend on x_i or y_{i+1}, is called a first order difference equation. If F does not depend on y_{i+1}, then (1.2) is said to be explicit. Otherwise, it is said to be implicit.

With difference equations now defined, it is natural to discuss the solutions of such equations.

Definition 1.5. A solution of (1.2) on a given R_{n+1} set is any function $y_i = y(x_i)$, $i = 0,1,\dots,n$, which satisfies (1.2) for $i = 0,1,2,\dots,n-1$.

It is of essence to understand that we are taking only discrete functions to be solutions of difference equations. This is reasonable since we have defined difference equations only on finite point sets. Indeed, it has been the unnatural attempt to find continuous solutions of difference equations which has led to the commonly held misconception that nonlinear difference equations are more difficult to solve than nonlinear differential equations.

Definition 1.6. An initial value (sometimes abbreviated I.V.) problem for (1.2) is one in which we must find a solution of (1.2) when y_0 is given.

Example. On the R_5 set $x_0 = 0$, $x_1 = 0.25$, $x_2 = 0.50$, $x_3 = 0.75$, $x_4 = 1.00$, the problem of finding a solution of

$$y_{i+1} = y_i^2 - 4x_i, \quad i = 0,1,2,3, \tag{1.3}$$

for which

$$y_0 = 0 \tag{1.4}$$

is an I.V. problem. Let us show how simple it is to generate the solution of this problem. Equation (1.3) is, in fact, four equations, namely,

$$y_1 = y_0^2 - 4x_0$$
$$y_2 = y_1^2 - 4x_1$$
$$y_3 = y_2^2 - 4x_2$$
$$y_4 = y_3^2 - 4x_3 .$$

From (1.4) and the known values of x_0, x_1, x_2, x_3, these equations yield, in order,

$$y_1 = 0^2 - 4(0) = 0$$

$$y_2 = 0^2 - 4(0.25) = -1$$

$$y_3 = (-1)^2 - 4(0.5) = -1$$

$$y_4 = (-1)^2 - 4(0.75) = -2 ,$$

so that the discrete function

$$y_0 = 0,\ y_1 = 0,\ y_2 = -1,\ y_3 = -1,\ y_4 = -2$$

is the solution of the given problem. If R_{n+1} were to consist of 10^6 points, we would have employed a computer to generate the solution.

The above discussion culminates in the following theorem, which assures, a priori, existence and uniqueness.

Theorem 1.1. If (1.2) is explicit, then an I.V. problem for (1.2) has a unique solution provided only that F is defined at each stage of iteration (1.2). The solution can be generated on a computer provided that F is a computer function and that no y_{i+1} is in absolute value larger than the largest number in the computer.

If (1.2) is implicit, then an existence and uniqueness theorem like the above is rarely easy to establish. Such matters will be discussed with each specific implicit equation, when it is introduced.

Finally, for convenience, we introduce the concept of stability, as follows.

Definition 1.7. The solution, or the computation of the solution, of an I.V. problem is said to be stable if no y_{i+1} is in absolute value larger than the largest number in one's computer.

Definition 1.7 provides a constructive form of the popular mathematical definition of stability, that sequence

y_{i+1} is stable if sequence $|y_{i+1}|$ is bounded, in that the bound is prescribed as the greatest number in one's computer.

Until we are actually confronted by stability problems, this section contains all the mathematics needed for the study of physical problems. Let us then turn, next, to very basic physics.

1.3 DISCRETE MECHANICS

Let us begin our study of physics by concentrating only on very elementary, one-dimensional, Newtonian motion. Our fundamental concepts, which, from a purely theoretical point of view can be left undefined (Russell), are particle, mass, force, distance, motion and time. From a heuristic point of view, however, these concepts can be described informally as follows.

A particle is some convenient, small spherical object, in three dimensions, or circular object, in two dimensions, of uniform weight, or mass, whose center is called its centroid. A particle's motion is described by the motion of its center. All larger bodies are made up of particles. Though force and distance can be thought of in the classical way (Synge and Griffith), motion will be thought of physiologically, as follows. Let Δx and Δt be positive numbers. On an X-axis, let $X_k = k\Delta x$, $k = 0,1,2,\ldots,m$, while on a T-axis, let $t_j = j\Delta t$, $j = 0,1,2,\ldots,n$, as shown in Figure 1.4. For illustrative purposes, assume that a particle P has center C at X_0 when $t = t_0$, at X_3 when $t = t_1$, at X_7 when $t = t_2$, at X_8 when $t = t_3$, and at X_6 when $t = t_4$. Then the motion of P from X_0 to X_6 is viewed merely as C's being at X_0, X_3, X_7, X_8, and X_6 at the respective times t_0, t_1, t_2, t_3, and t_4.

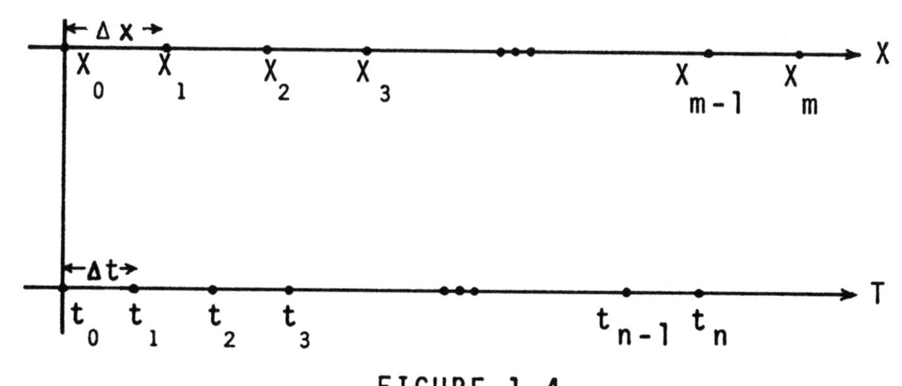

FIGURE 1.4

Thus, P's motion is conceived of as a sequence of "stills," which is, of course, biologically acceptable and realized in motion pictures, where motion is observed from a finite sequence of stills projected with sufficient rapidity and transmitted as retinal images to the brain. The passage of time can be thought of in terms of the uniform motion of a time particle of mass Δt.

In order to formulate, next, the concepts of velocity and acceleration, let us consider a simple falling body experiment and see what it suggests. For $t_k = k\Delta t$, $k = 0,1,2,\ldots,n$, consider a particle which is dropped vertically from a position of rest. Let the distance fallen at time t_k be x_k. Denote the particle's velocity and acceleration at time t_k by v_k and a_k, respectively. Since the initial position is one of rest, set $v_0 = 0$, and focus attention first on v_1. Motivated by the practical values of both symmetry and smoothing, let us measure v_1 by

$$\frac{v_1 + v_0}{2} = \frac{x_1 - x_0}{\Delta t} ,$$

which, in turn, motivates the general formula

$$(1.5) \qquad \frac{v_{k+1} + v_k}{2} = \frac{x_{k+1} - x_k}{\Delta t} , \quad k = 0,1,2,\ldots,n-1 .$$

Similar reasoning, however, cannot be applied to acceleration, for, in general, unless one has some additional knowledge about the forces acting upon the particle, a_0 is not known. Hence, we merely set

$$a_0 = \frac{v_1 - v_0}{\Delta t},$$

which, in turn, motivates the general formula

$$(1.6) \qquad a_k = \frac{v_{k+1} - v_k}{\Delta t}, \quad k = 0,1,2,\ldots,n-1 .$$

For any particle, then, whose motion is one dimensional, let $\Delta t > 0$ and $t_k = k\Delta t$, $k = 0,1,2,\ldots,n$. If the particle at time t_k is located at x_k, then its velocity $v(t_k) = v_k$ is defined as the rate of change of distance with respect to time as measured by (1.5), while its acceleration is defined as the rate of change of velocity with respect to time as measured by (1.6).

With regard to (1.5), observe that

$$\begin{aligned}
v_1 &= \frac{2}{\Delta t}[x_1-x_0]-v_0 \\
v_2 &= \frac{2}{\Delta t}[x_2-x_1]-v_1 = \frac{2}{\Delta t}[x_2-2x_1+x_0]+v_0 \\
v_3 &= \frac{2}{\Delta t}[x_3-x_2]-v_2 = \frac{2}{\Delta t}[x_3-2x_2+2x_1-x_0]-v_0 \\
v_4 &= \frac{2}{\Delta t}[x_4-x_3]-v_3 = \frac{2}{\Delta t}[x_4-2x_3+2x_2-2x_1+x_0]+v_0, \\
&\vdots
\end{aligned}$$

which motivates the following theorem.

__Theorem 1.2.__ Formula (1.5) implies that

$$v_1 = \frac{2}{\Delta t}[x_1-x_0]-v_0$$

$$v_k = \frac{2}{\Delta t}[x_k+(-1)^k x_0 + 2\sum_{j=1}^{k-1}(-1)^j x_{k-j}]+(-1)^k v_0; \ k\geq 2.$$

Proof. The proof follows immediately by mathematical induction.

In an analogous fashion, study of $a_1, a_2, a_3, a_4, a_5, \ldots,$ motivates the following theorem.

Theorem 1.3. If a_k is defined by (1.6), then

$$a_0 = \frac{2}{(\Delta t)^2}[x_1 - x_0 - v_0\Delta t]$$

$$a_1 = \frac{2}{(\Delta t)^2}[x_2 - 3x_1 + 2x_0 + v_0\Delta t]$$

$$a_{k-1} = \frac{2}{(\Delta t)^2}\{x_k - 3x_{k-1} + 2(-1)^k x_0 + 4\sum_{j=2}^{k-1}[(-1)^j x_{k-j}] + (-1)^k v_0\Delta t\}, \quad k\ge 3.$$

Proof. From (1.6) and Theorem 1.2,

$$a_0 = \frac{v_1 - v_0}{\Delta t} = \frac{1}{\Delta t}\{\frac{2}{\Delta t}(x_1 - x_0) - 2v_0\} = \frac{2}{(\Delta t)^2}[x_1 - x_0 - v_0\Delta t]$$

$$a_1 = \frac{v_2 - v_1}{\Delta t} = \frac{1}{\Delta t}\{[\frac{2}{\Delta t}(x_2 - 2x_1 + x_0) + v_0] - [\frac{2}{\Delta t}(x_1 - x_0) - v_0]\}$$

$$= \frac{2}{(\Delta t)^2}[x_2 - 3x_1 + 2x_0 + v_0\Delta t],$$

while, for $k\ge 3$,

$$a_{k-1} = \frac{v_k - v_{k-1}}{\Delta t}$$

$$= \frac{1}{\Delta t}\{\frac{2}{\Delta t}[x_k + (-1)^k x_0 + 2\sum_{j=1}^{k-1}(-1)^j x_{k-j}] + (-1)^k v_0$$

$$- \frac{2}{\Delta t}[x_{k-1} + (-1)^{k-1}x_0 + 2\sum_{j=1}^{k-2}(-1)^j x_{k-j-1}]$$

$$- (-1)^{k-1}v_0\}$$

$$= \frac{2}{(\Delta t)^2}\{x_k - 3x_{k-1} + 2(-1)^k x_0 + 4\sum_{j=2}^{k-1}[(-1)^j x_{k-j}]$$

$$+ (-1)^k v_0\Delta t\},$$

and the theorem is proved.

Finally, let us replace Newton's classical dynamical differential equation by a difference equation. We will call the replacement a discrete Newton's equation and, for the present, take it in the particularly simple form

$$(1.7) \quad ma_k = F(x_k, v_k, v_{k+1}), \quad k = 0,1,2,\ldots,n-1.$$

With regard to (1.7), note that the exact structure of F will depend on the problem under consideration, while (1.7) is, by (1.6), a first order difference equation in v. Note also that, if F is independent of v_{k+1}, then, from Theorems 1.2 and 1.3, the left side of (1.7) is a linear combination of $x_0, x_1, x_2, \ldots, x_k, x_{k+1}$, while the right side is a function only of $x_0, x_1, x_2, \ldots, x_k$. Thus, whether F in (1.7) is linear or not, the equation can be solved for x_{k+1} to yield an explicit recursion formula for the sequence of particle positions. Though this provides us with a constructive method for using a computer to generate the particle's motion, we will show later how to do it even more simply. For the present, however, let us turn to the matter of conservation principles.

1.4 CONSERVATION OF ENERGY AND MOMENTUM

We will show now, interestingly enough, that the completely arithmetic, discrete mechanics formulation of Section 1.3 yields various conservation principles of classical mechanics. This is not valid for traditional discretizations (see, e.g., Greenspan (6), Henrici, Moore, and Richtmyer and Morton) of continuous mechanics. The impact upon the stability of one's calculations will be explored in the next chapter.

For $\Delta t > 0$ and $t_k = k\Delta t$, $k = 0,1,2,\ldots,n$, let particle P at time t_k be at x_k, have velocity v_k, and have acceleration a_k. Further, let the motion of P be determined by the Newtonian difference equation (1.7). Then the work W done by F on P from t_0 to t_n is defined by

$$(1.8) \qquad W = \sum_{i=0}^{n-1} (x_{i+1} - x_i)F_i ,$$

where $F_i \equiv F(x_i, v_i, v_{i+1})$. Then, from (1.5)-(1.7),

$$\begin{aligned} W &= \sum_{0}^{n-1} (x_{i+1} - x_i)ma_i \\ &= m \sum_{0}^{n-1} (x_{i+1} - x_i)\left(\frac{v_{i+1} - v_i}{\Delta t}\right) \\ &= m \sum_{0}^{n-1} \left(\frac{v_{i+1} + v_i}{2}\right)(v_{i+1} - v_i) \\ &= \frac{m}{2} \sum_{0}^{n-1} (v_{i+1}^2 - v_i^2) \\ &= \frac{m}{2} v_n^2 - \frac{m}{2} v_0^2 . \end{aligned}$$

Defining the kinetic energy K_i at time t_i by

$$(1.9) \qquad K_i = \frac{1}{2} mv_i^2$$

yields, then, the classical relationship

$$(1.10) \qquad W = K_n - K_0 .$$

Note now that (1.10) is valid independently of the specific structure of F. In order to proceed to the study of energy conservation, one must have in addition to kinetic energy the concept of potential energy. The concept of potential energy, however, varies with the particular form of

F, so that we must now be very specific about the kind of force acting on the particle. For the present, we will consider only the force of gravity. . Later, when considering, for example, oscillators and planetary motions, other forces will be considered.

Consider again the falling body experiment described in Section 1.3. The work done when the body has fallen from x_0 to x_n is still given by (1.10). However, one can now find a second formula for W as follows. Assuming that the only force acting upon the particle is gravity and that this force is constant, let

$$(1.11)\quad F = -mg\ .$$

Then, from (1.8) and (1.11)

$$\begin{aligned} W &= \sum_{i=0}^{n-1} (x_{i+1} - x_i)(-mg) \\ &= \sum_{0}^{n-1} (-mgx_{i+1} + mgx_i) \\ &= -mgx_n + mgx_0\ . \end{aligned}$$

If one defines the potential energy V_i at time t_i by

$$V_i = mgx_i\ ,$$

then

$$(1.12)\quad W = -V_n + V_0\ .$$

Finally, elimination of W between (1.10) and (1.12) implies

$$K_n - K_0 = -V_n + V_0\ ,$$

or, equivalently,

(1.13) $K_n + V_n = K_0 + V_0$,

which is the classical law of conservation of energy. Another way of stating (1.13) is that since n is arbitrary, the sum $K_i + V_i$ is the same for $i = 0,1,2,\ldots,n$, that is, it is a time invariant.

Finally, as shown in Figure 1.5, consider two particles in motion. Let the first have center A, mass m_A and velocity $v(A)$, while the second has center B, mass m_B and velocity $v(B)$. We consider a relative motion in which there is an elastic collision. During the collision, let

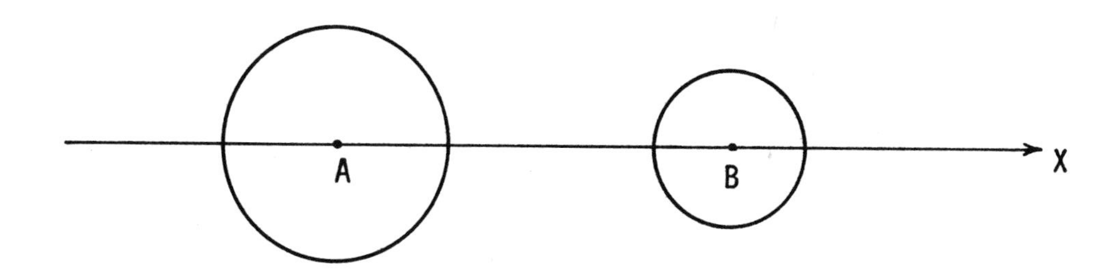

FIGURE 1.5

$F(A)$ be the force of the second body on the first while $F(B)$ is the force of the first body on the second. These forces are non-constant and satisfy

(1.14) $F(A) = -F(B)$.

Now, at each time t_i during the contact period,

$$F_i(A) = m_A a_i(A) ,$$

so that

$$\sum_{i=0}^{n-1} F_i(A)\Delta t = \sum_{i=0}^{n-1} m_A a_i(A)\Delta t$$

$$= m_A \sum_{i=0}^{n-1} [v_{i+1}(A) - v_i(A)]$$

$$= m_A[v_n(A) - v_0(A)] .$$

Similarly,

$$\sum_{i=0}^{n-1} F_i(B)\Delta t = m_B[v_n(B) - v_0(B)] .$$

But, from (1.14),

$$\sum_{i=0}^{n-1} F_i(A)\Delta t = - \sum_{i=0}^{n-1} F_i(B)\Delta t ,$$

so that

$$m_A[v_n(A) - v_0(A)] = -m_B[v_n(B) - v_0(B)] ,$$

or, equivalently,

$$(1.15) \quad m_A v_n(A) + m_B v_n(B) = m_A v_0(A) + m_B v_0(B).$$

If one defines the momentum M_i at time t_i of a particle of mass m by

$$M_i = mv_i ,$$

then (1.15) is the classical law of conservation of linear momentum.

1.5 REMARKS

It should be noted that, though certain major laws of Newtonian mechanics are now established for discrete mechanics, there are some minor results in continuous mechanics which are not valid in discrete mechanics. For example, if a particle's motion under a force F begins and ends at the same point, the work done by F need not, as defined by (1.8), be zero.

Note also that, from the definition $V_i = mgx_i$ of potential energy in Section 1.4, one has that

$$\frac{V_{i+1} - V_i}{x_{i+1} - x_i} = mg ,$$

or, in a more concise form,

$$\frac{\Delta V_i}{\Delta x_i} = -F , \tag{1.16}$$

where Δ is a forward difference operator defined for any function W by $\Delta W_i = W_{i+1} - W_i$. The relationship (1.16) is the difference analogue of the classical relationship between force and potential energy.

Finally, note that if one uses a ruler and a clock, with respective accuracies Δx and Δt in a given physical experiment, then taking these values in the dynamical equations of a discrete model of the experiment actually corresponds to including the measuring instruments as part of the model (Greenspan (6)).

CHAPTER II - DISCRETE OSCILLATORS

2.1 INTRODUCTION

In this chapter we will study the simplest, nontrivial motion which is in one direction only, namely, oscillation. Intuitively, oscillation is motion which is back and forth over all, or part, of a one dimensional path. Examples of oscillators are electrons in an atom, vibrating springs, and, in appropriate coordinates, pendula (Feynman, Leighton, and Sands).

2.2 DAMPED MOTION IN A NONLINEAR FORCE FIELD

Because the required analysis is typical, and because of its importance in classical mechanics, attention will be directed to the study of damped, oscillatory motion in a nonlinear force field. Such motion is, classically, nonconservative. The discussion will serve also to illustrate the computer implementation of the methods and theory of Chapter I when the dynamical equation is explicit.

As shown in Figure 2.1, consider a particle of unit mass which is constrained to move with its center C in an X direction. A displacement of the particle such that the directed distance OC is x_i is, for illustrative purposes, assumed to be opposed by a field force of magnitude $\sin x_i$ and by a viscous damping force of magnitude αv_i, where α is a positive constant. Such a set of interacting forces is

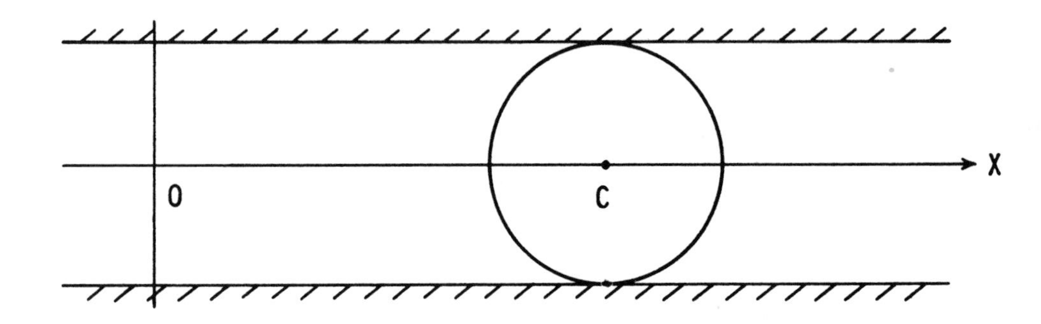

FIGURE 2.1

typical in the analysis of the motion of a pendulum (Greenspan (6)). Then the equation of motion takes the particular form

$$a_k = -\alpha v_k - \sin x_k, \quad k = 0,1,2,\ldots,n-1 . \tag{2.1}$$

But, from (1.5) and (1.6),

$$v_{k+1} = v_k + a_k \Delta t, \quad k = 0,1,2,\ldots,n-1 \tag{2.2}$$

and

$$x_{k+1} = x_k + \frac{\Delta t}{2} (v_{k+1} + v_k), \quad k = 0,1,2,\ldots,n-1, \tag{2.3}$$

so that the motion of C can be generated recursively as follows. Fix x_0 and v_0, that is, an initial position and an initial velocity, respectively. Generate a_0 from (2.1), then v_1 from (2.2), and finally position x_1 from (2.3). Next, using x_1 and v_1, calculate a_1 from (2.1), then v_2 from (2.2) and finally x_2 from (2.3). In the indicated fashion, from x_k and v_k, generate a_k from (2.1), then v_{k+1} from (2.2), and finally x_{k+1} from (2.3). Since, for any x_k and v_k, (2.1)-(2.3) imply that a_k, v_{k+1} and x_{k+1} exist and are unique, it follows immediately from the above discussion that the motion of C is uniquely defined once initial conditions x_0 and v_0 are given, or, in more general terminology, the solution of an initial value problem for (2.1) exists and is unique. The question of stability, however, is somewhat delicate, and will be studied in the next section.

As an illustrative example, the solution of (2.1)-(2.3) with the parameter values $\alpha = 0.3$, $\Delta t = 0.01$, $x_0 = \pi/4$, $v_0 = 0$, $n = 15000$, was generated on the UNIVAC 1108 in under 30 seconds. The portion of the curve between t_0 and t_{2500} is shown in Figure 2.2, which exhibits strong damping and peak

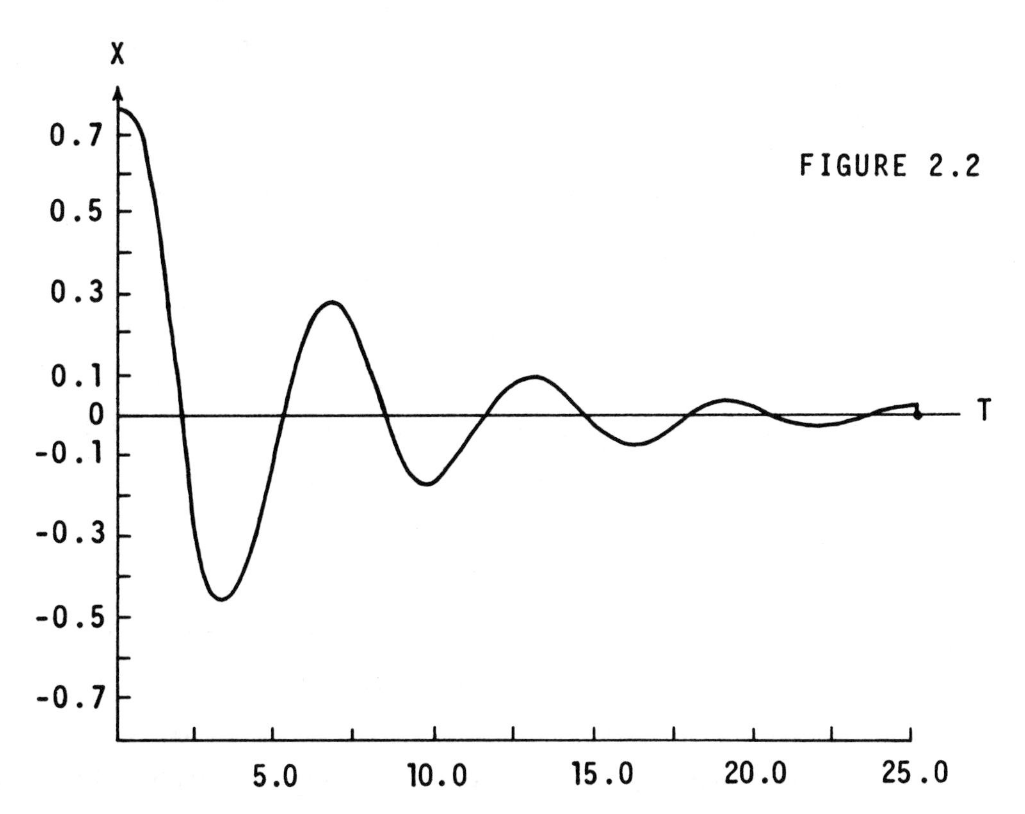

FIGURE 2.2

values of 0.754615, -0.480136, 0.297837, -0.185740, 0.116061, -0.072585, 0.045407, and -0.028421 at the respective times 0, 3.27, 6.48, 9.66, 12.84, 16.01, 19.18 and 22.35. The time required for the particle to travel from one peak successively to the next decreased monotonically, which is consistent with physical experimentation.

2.3 STABILITY

The computation defined by (2.1)-(2.3) is, in every sense, physically realistic, since damping and nonlinear force effects are believed to be present, to a greater or lesser degree, in all physical systems. Let us examine, then, ways in which one can study the stability of the system.

Definition 1.6 sets forth a criterion by which one determines stability from computer output. To use such a method wisely, one must also use physical common sense. For example, if a pendulum has very little damping and is expected to swing appreciably for at least 2 hours, then computer output which is stable for the first 10 seconds of pendulum oscillation would be meaningless. Indeed, to each physical problem one should associate a fixed time period during which significant motion can be expected and one should check that one has stability over this time period, at least. Using such a method, it was determined (Greenspan (6)) experimentally that (2.1)-(2.3) is stable on the UNIVAC 1108 if

$$(2.4) \qquad \Delta t < 2\alpha \ .$$

A stability condition like (2.4) would be invaluable if it were available <u>before</u> one began to compute, and herein lies the essential value of the mathematical analysis of stability, when it exists. Unfortunately, <u>general</u> mathematical methods for stability analysis are available only for linear problems (Henrici), so that if such an analysis were to exist for the nonlinear problem at hand, one would expect it to be full of very special tricks. Interestingly enough, such an analysis does exist (Cryer) and will be described next. Note that though, in reality, one terminates the iteration (2.1)-(2.3) in a finite time and has to contend with roundoff errors in the calculation, the mathematical approach is to consider (2.1)-(2.3) as $n \to \infty$ and to show, only, that, neglecting roundoff error, the sequences so generated are bounded.

First, let us write system (2.1)-(2.3) in the equivalent form

$$(2.5)\qquad x_{k+1} - x_k = \frac{\Delta t}{2}(v_k + v_{k+1})$$

$$(2.6)\qquad v_{k+1} - v_k = -\Delta t\,(\alpha v_k + \sin x_k).$$

Multiplication of (2.6) by $(v_{k+1} + v_k)/2$ implies

$$\frac{v_{k+1}^2 - v_k^2}{2} = -\frac{\Delta t}{2}(v_{k+1} + v_k)(\alpha v_k + \sin x_k),$$

which, from (2.5), implies

$$(2.7)\qquad \frac{v_{k+1}^2 - v_k^2}{2} = -\frac{\alpha\Delta t}{2}(v_{k+1}+v_k)v_k - (x_{k+1}-x_k)\sin x_k.$$

But,

$$-(\cos x_{k+1} - \cos x_k) = (x_{k+1}-x_k)\sin x_k + \frac{1}{2}(x_{k+1}-x_k)^2\cos \bar{x}_k,$$

where $\bar{x}_k$ is between x_k and x_{k+1}. Hence,

$$(2.8)\qquad -(\cos x_{k+1} - \cos x_k) \le (x_{k+1}-x_k)\sin x_k + \frac{(x_{k+1}-x_k)^2}{2}.$$

Adding (2.7) and (2.8) then yields

$$\frac{v_{k+1}^2 - v_k^2}{2} - (\cos x_{k+1} - \cos x_k)$$
$$\leq -\frac{\alpha\Delta t}{2}(v_{k+1}+v_k)v_k + \frac{(x_{k+1}-x_k)^2}{2}.$$

With the aid of (2.5), one has

$$\frac{v_{k+1}^2 - v_k^2}{2} - (\cos x_{k+1} - \cos x_k)$$
$$\leq -\frac{\alpha\Delta t}{2}(v_{k+1}+v_k)v_k + \frac{(v_{k+1}+v_k)^2}{8}(\Delta t)^2$$
$$= \frac{\Delta t}{8}[(\Delta t)v_{k+1}^2 + (-4\alpha+\Delta t)v_k^2 + (-4\alpha+2\Delta t)v_k v_{k+1}]$$
$$= \frac{\Delta t}{8}[(\Delta t-2\alpha)(v_{k+1}+v_k)^2 + 2\alpha(v_{k+1}^2 - v_k^2)],$$

from which it follows that

$$\text{(2.9)}\quad \left(\frac{v_{k+1}^2 - v_k^2}{2}\right)\left(1 - \frac{\alpha\Delta t}{2}\right) - (\cos x_{k+1} - \cos x_k)$$
$$\leq \frac{\Delta t}{8}(\Delta t-2\alpha)(v_{k+1}+v_k)^2.$$

Setting

$$\text{(2.10)}\quad E_k = \frac{v_k^2}{2}\left(1 - \frac{\alpha\Delta t}{2}\right) - \cos x_k,$$

one can rewrite (2.9) as

$$\text{(2.11)}\quad E_{k+1} - E_k \leq \frac{\Delta t}{8}(\Delta t-2\alpha)(v_{k+1}+v_k)^2.$$

Summing both sides of (2.11) from zero to $n-1$ implies

$$\text{(2.12)}\quad E_n - E_0 \leq \frac{\Delta t}{8}(\Delta t-2\alpha)\sum_{k=0}^{n-1}(v_{k+1}+v_k)^2.$$

Setting

$$(2.13)\quad F_n = \frac{\Delta t}{8}(2\alpha-\Delta t)\sum_{k=0}^{n-1}(v_{k+1}+v_k)^2$$

allows one to write

$$(2.14)\quad E_n + F_n \le E_0, \quad n\ge 1.$$

To continue, assume with regard to (2.10) and (2.12) that

$$(1-\frac{\alpha\Delta t}{2}) > 0, \quad (2\alpha-\Delta t) > 0\ ,$$

or, equivalently, that

$$(2.15)\quad \Delta t < \min\left[2\alpha, \frac{2}{\alpha}\right]\ .$$

Several interesting consequences now follow. From (2.10) and (2.15),

$$(2.16)\quad -1 \le E_k, \quad k = 0,1,2,\ldots,n,$$

so that (2.13)-(2.16) imply

$$(2.17)\quad 0 \le F_n \le E_0 + 1, \quad n\ge 1\ .$$

Hence, sequence F_n is bounded. Also, since $F_n \ge 0$, it follows from (2.14) and (2.16) that

$$(2.18)\quad E_n = \frac{v_n^2}{2}(1-\frac{\alpha\Delta t}{2}) - \cos x_n \le E_0, \quad n\ge 1.$$

From (2.16) and (2.18), then, sequence E_k is bounded and

$$(2.19)\quad |v_n| \le \frac{2(|E_0|+2)^{\frac{1}{2}}}{(2-\alpha\Delta t)^{\frac{1}{2}}}\ , \quad n\ge 1.$$

Hence, <u>the sequence</u> v_k <u>is bounded</u>.

Next, observe that, from (2.13), F_n is monotonic increasing and, since it is bounded above,

$$(2.20)\quad |v_{k+1} + v_k| \to 0 .$$

From (2.5), then,

$$(2.21)\quad |x_{k+1} - x_k| \to 0 .$$

However, (2.6) implies

$$(2.22)\quad v_{k+2} - v_{k+1} = -\Delta t(\alpha v_{k+1} + \sin x_{k+1}) ,$$

which, when added to (2.6), gives

$$(2.23)\quad (v_{k+2} + v_{k+1})-(v_{k+1} + v_k) = -\Delta t[\alpha(v_{k+1} + v_k) + (\sin x_{k+1} + \sin x_k)].$$

From (2.20) and (2.23), then,

$$(\sin x_{k+1} + \sin x_k) \to 0 ,$$

which, together with (2.21) implies that $\sin x_k \to 0$. Hence, from (2.21) sequence x_k converges to $K\pi$, where K is a fixed integer. Thus, <u>sequence</u> x_k <u>is bounded</u>.

Finally, note that <u>sequence</u> a_k, defined by (2.1), <u>is also bounded</u>, since each of v_k and x_k is. Thus, the following theorem has been proved.

<u>Theorem 2.1</u>. If α is a fixed, positive damping factor and Δt satisfies

$$\Delta t < \min [2\alpha, 2/\alpha],$$

then the sequences x_k, v_k, and a_k, generated recursively by (2.1)-(2.3) from given initial data x_0 and v_0 are bounded.

The criterion (2.4) and that of Theorem 2.1 are equivalent in the physically important case $\alpha \le 1$. It is then unfortunate that such mathematical analysis is so difficult to develop.

2.4 THE HARMONIC OSCILLATOR

Any oscillator which is periodic and free from damping, and is therefore conservative, is called a harmonic oscillator. It is believed that there is no natural harmonic oscillator. Nevertheless, in certain fields of study, like plasma dynamics (Symon, Marshall, and Li), the accuracy constraints are sufficiently large at the present time so as to allow certain motions to be considered as harmonic. The associated dynamical differential equation is then of the particularly simple form

$$(2.24) \quad m\ddot{x} + \omega^2 x = 0 .$$

Linear differential equation (2.24) can be solved completely by analytical methods so that recourse to discretization is unnecessary. Let us then merely show that a discrete, conservative model of the harmonic oscillator does exist.

For $\Delta t > 0$ and $t_k = k\Delta t$, $k = 0,1,2,\ldots,n$, let particle P of mass m be at x_k at time t_k. Then P is said to be a harmonic oscillator if its motion is determined by the discrete, dynamical equation

$$(2.25) \quad ma_k + \omega^2 \frac{x_{k+1} + x_k}{2} = 0 ,$$

where ω is a nonzero constant. Let us set

$$(2.26) \quad F_k = -\omega^2 \frac{x_{k+1} + x_k}{2} .$$

Then, from formula (1.8) for work, one has

$$W = \sum_{0}^{n-1} (x_{i+1} - x_i)\left(-\omega^2 \frac{x_{i+1} + x_i}{2}\right)$$

$$= -\frac{\omega^2}{2} x_n^2 + \frac{\omega^2}{2} x_0^2 .$$

If the potential energy V_k at t_k is defined by

$$(2.27) \quad V_k = \frac{\omega^2}{2} x_k^2 ,$$

then

$$(2.28) \quad W = -V_n + V_0 .$$

Thus, from (1.10), which was independent of the form of F, one has

$$K_n - K_0 = -V_n + V_0,$$

or, equivalently,

$$(2.29) \quad K_n + V_n = K_0 + V_0 ,$$

which is, of course, a statement of the conservation of energy. Note that, as indicated previously, it is the proper choice of a potential function, which this time is (2.27), which leads to (2.29), and, as in the case of gravity, it is the telescopic nature of the summation in the formula for work that leads directly to the desired result.

2.5 REMARKS

Note that (1.5) and (1.7) imply that (2.25) is implicit. Nevertheless, because the equation is linear, it can be rewritten in an explicit form (Greenspan (11)), so that existence, uniqueness, computability, and stability for any initial value problem follow readily. However, since no computation will be done with (2.26), the analysis will not be given here.

CHAPTER III - NONLINEAR STRING VIBRATIONS

3.1 INTRODUCTION

After having discussed the motion of a single oscillator, it is natural to consider the motion of a system of oscillators. Since the dynamics of a vibrating string make it especially suitable for such a study, it is to such vibrations that the present chapter is directed.

3.2 THE DISCRETE STRING

A discrete string is one which is composed of a finite number of particles. It will be treated mathematically as an ordered set of $m+2$ homogeneous particles P_k, $k = 0,1,\ldots,m,m+1$, with respective centers (x_k, y_k), as shown typically in Figure 3.1.

Our problem will be that of describing the return of a discrete string to a position of equilibrium from an initial position of tension. Due to current technological limitations, the number of particles will be relatively small. The resulting motion is considered to be an approximation to that of a real string, the improvement of which depends largely on one's computer capability. It will be assumed through-

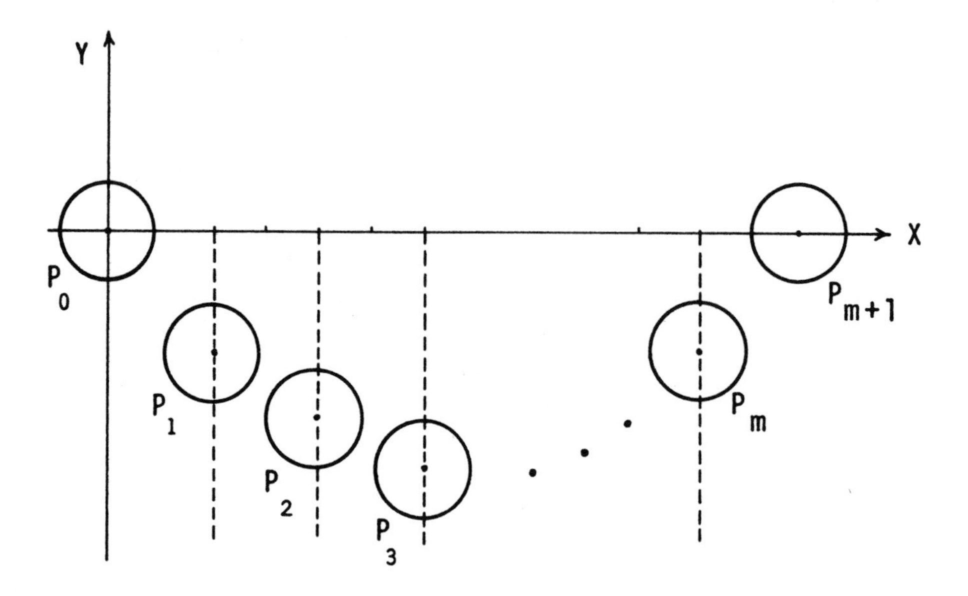

FIGURE 3.1

out that P_0 and P_{m+1} are fixed while $P_1, P_2, \ldots, P_m$ are free to move, and that

$$(3.1) \qquad x_0 = y_0 = y_{m+1} = 0 \ .$$

3.3 EQUATIONS OF DISCRETE STRING VIBRATION

We proceed under the popular assumption that horizontal particle motion is negligible, so that each particle of a discrete string can move in the vertical direction only. The string is said then to exhibit transverse vibrations.

For $\Delta x > 0$, let $x_j = j(\Delta x)$, $j = 0,1,2,\ldots,m+1$, be an R_{m+2} set. For $\Delta t > 0$, let time $t_k = k\Delta t$, $k = 0,1,2,\ldots,n$, be measured in seconds, and let P_j, as shown in Figure 3.2, be a typical particle in motion. In order to incorporate the time dependence of the centers of P_{j-1}, P_j, and P_{j+1}, let the respective centers of these particles at time t_k be $(x_{j-1}, y_{j-1,k})$, $(x_j, y_{j,k})$, $(x_{j+1}, y_{j+1,k})$, where each coordinate is measured in feet. Also, let $a_{j,k}$ and $v_{j,k}$ be the acceleration and velocity, respectively, of P_j at time t_k.

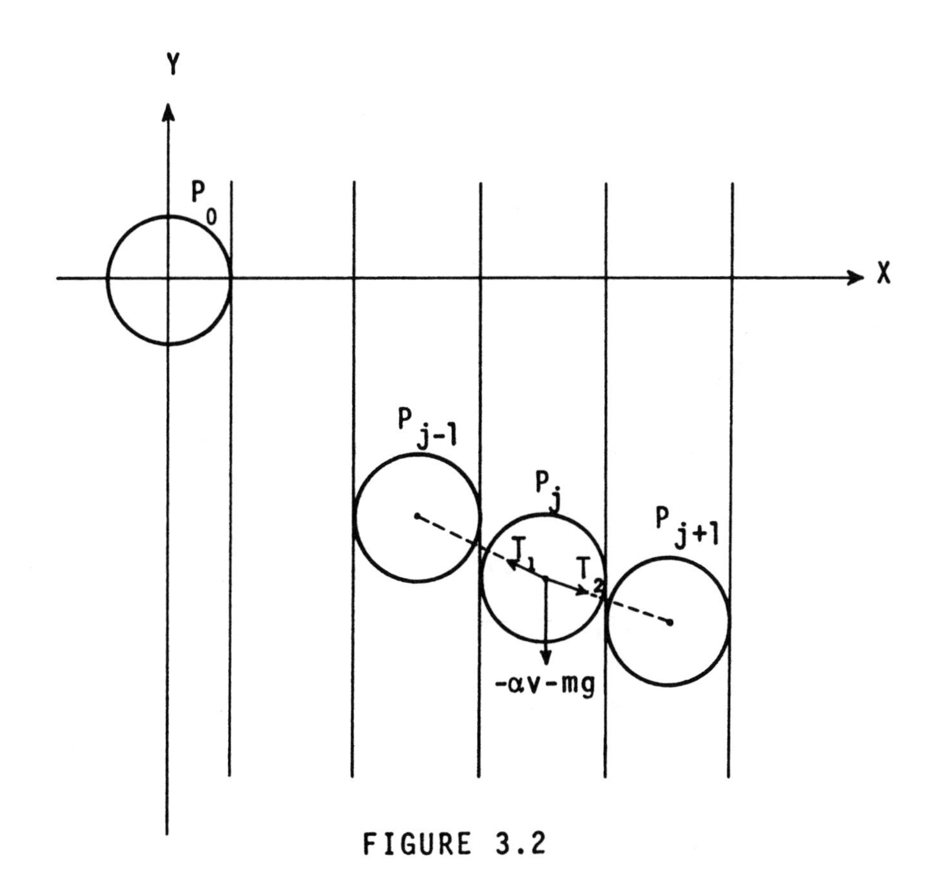

FIGURE 3.2

In studying the motion of each P_j, we will consider only tensile, viscous, and gravitational forces. For this purpose, let T_1 be the tensile force between P_{j-1} and P_j, let T_2 be the tensile force between P_j and P_{j+1}, and let the viscosity vary with the velocity of the particle. Then, for each particle P_j, the discrete Newtonian equation which determines its motion is

$$(3.2)\quad \bar{m}a_{j,k} = |T_2| \frac{y_{j+1,k}-y_{j,k}}{[(\Delta x)^2+(y_{j+1,k}-y_{j,k})^2]^{1/2}}$$

$$- |T_1| \frac{y_{j,k}-y_{j-1,k}}{[(\Delta x)^2+(y_{j,k}-y_{j-1,k})^2]^{1/2}}$$

$$- \alpha v_{j,k} - \bar{m}g; \quad j=1,2,\ldots,m; \quad k=0,1,2,\ldots,n-1,$$

where $g = 32.2$, $\alpha \geq 0$, and $\bar{m}$ is the mass of each P_j. Since $v_{j,k}$ and $a_{j,k}$ are in the Y direction only, we have, in analogy with (2.2) and (2.3),

$$(3.3)\quad v_{j,k+1} = v_{j,k} + a_{j,k}\Delta t; \; j=1,2,\ldots,m; \; k=0,1,2,\ldots,n-1$$

$$(3.4)\quad y_{j,k+1} = y_{j,k} + \frac{\Delta t}{2}(v_{j,k+1}+v_{j,k}); \quad j=1,2,\ldots,m;$$

$$k=0,1,2,\ldots,n-1 .$$

The precise steps, or, recipe, or, in computer terms, the algorithm, for generating the motion of a discrete string can be given now as follows. From prescribed initial particle positions $y_{j,0}$ and velocities $v_{j,0}$, $j = 1,2,\ldots,m$, one determines initial accelerations $a_{j,0}$, $j = 1,2,\ldots,m$, from (3.2). The results are used in (3.3) to determine the next velocities $v_{j,1}$ of the particles, and these velocities, in turn, are used in (3.4) to determine the corresponding positions $y_{j,1}$. To proceed to the second time step, the data $y_{j,1}$ and $v_{j,1}$ are used in (3.2) to determine the $a_{j,1}$, from which one determines the $v_{j,2}$ by means of (3.3), and then the $y_{j,2}$ by means of (3.4). In general, one steps ahead in time from $y_{j,k}$ and $v_{j,k}$ by determining the $a_{j,k}$ from (3.2), then the $v_{j,k+1}$ from (3.3), and finally the

$y_{j,k+1}$ from (3.4). The computational procedure then is completely analogous to that given for (2.1)-(2.3) except that this time we have to determine the motions of m particles, rather than one particle, at each time step.

Before considering actual dynamical problems, note that it is of computational advantage to know, a priori, the steady state, or equilibrium position, of a vibrating string problem. Physically, a vibrating string would converge to a steady state position, so that we require the same behavior from our computations. This steady state configuration can be obtained by applying the generalized Newton's method (Greenspan (6)) to the algebraic system

$$(3.5)\qquad \frac{|T_2|(y_{j+1}-y_j)}{[(\Delta x)^2+(y_{j+1}-y_j)^2]^{\frac{1}{2}}} - \frac{|T_1|(y_j-y_{j-1})}{[(\Delta x)^2+(y_j-y_{j-1})^2]^{\frac{1}{2}}}$$

$$= \bar{m}g; \quad j = 1,2,\ldots,m,$$

which results from (3.2) by setting $a_{j,k} \equiv v_{j,k} \equiv 0$.

3.4 EXAMPLES

From the large number of examples run on the UNIVAC 1108, we will discuss now several which are typical and illustrative. In all cases, the output is given graphically with 100 additional points interpolated linearly between each pair of consecutive particles. All strings to be considered are relatively heavy and are, approximately, of the same weight.

Example 1. Consider a twenty-one particle string with $x_j = \frac{j}{10}$, $j = 0,1,2,\ldots,20$; with T_1 and T_2 defined by

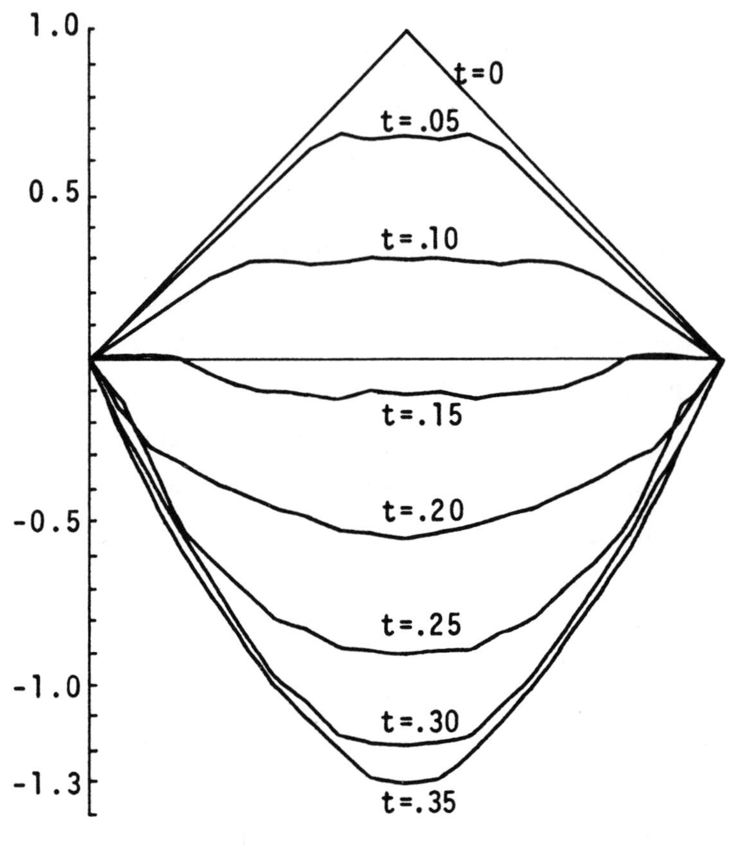

FIGURE 3.3

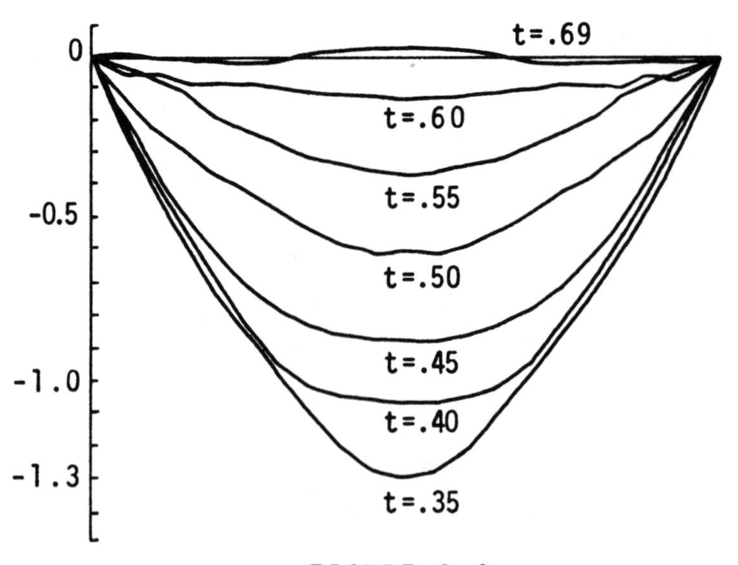

FIGURE 3.4

$$(3.6) \quad T_1 = T_0 \left[1 + \left|\frac{y_{j,k}-y_{j-1,k}}{\Delta x}\right| + \frac{\varepsilon}{2}\left(\frac{y_{j,k}-y_{j-1,k}}{\Delta x}\right)^2\right]$$

$$(3.7) \quad T_2 = T_0 \left[1 + \left|\frac{y_{j+1,k}-y_{j,k}}{\Delta x}\right| + \frac{\varepsilon}{2}\left(\frac{y_{j+1,k}-y_{j,k}}{\Delta x}\right)^2\right]$$

and with $\alpha = 0.15$, $\bar{m} = 0.05$, $T_0 = 12.5$, $\Delta t = 0.00025$, $\Delta x = 0.1$, $m = 19$, $\varepsilon = 0.01$. Formulas (3.6) and (3.7) are simple nonlinear relationships which describe the tension between successive particles as a function of the slope of the segment joining the centers of these particles. The string is placed in a position of tension by bringing the center particle to (1,1), while those to the left of the center are positioned on $y = x$ and those to the right of the center are positioned on $y = -x + 2$. The resulting configuration is that shown for $t = 0$ in Figure 3.3. The string is released from this position of tension, so that the initial velocity of each particle is zero. Its downward motion from $t = 0$ to $t = 0.35$ is shown typically in Figure 3.3, while its upward motion from $t = 0.35$ to $t = 0.69$ is shown typically in Figure 3.4. The lower curve in Figure 3.5 is the string's position after six seconds, at which time its maximum oscillation is less than 0.005. The upper curve in Figure 3.5, labelled S, is the steady state solution, which

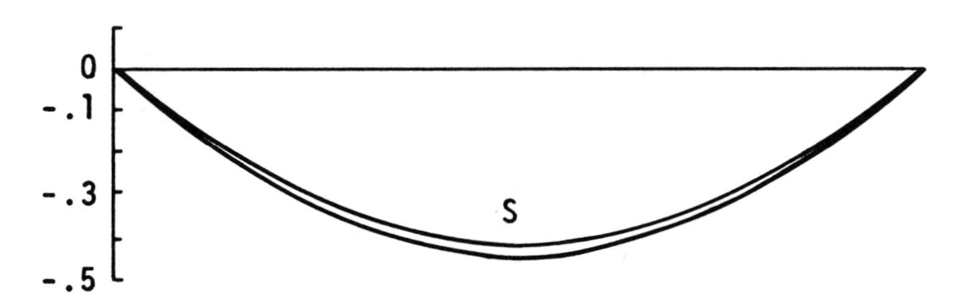

FIGURE 3.5

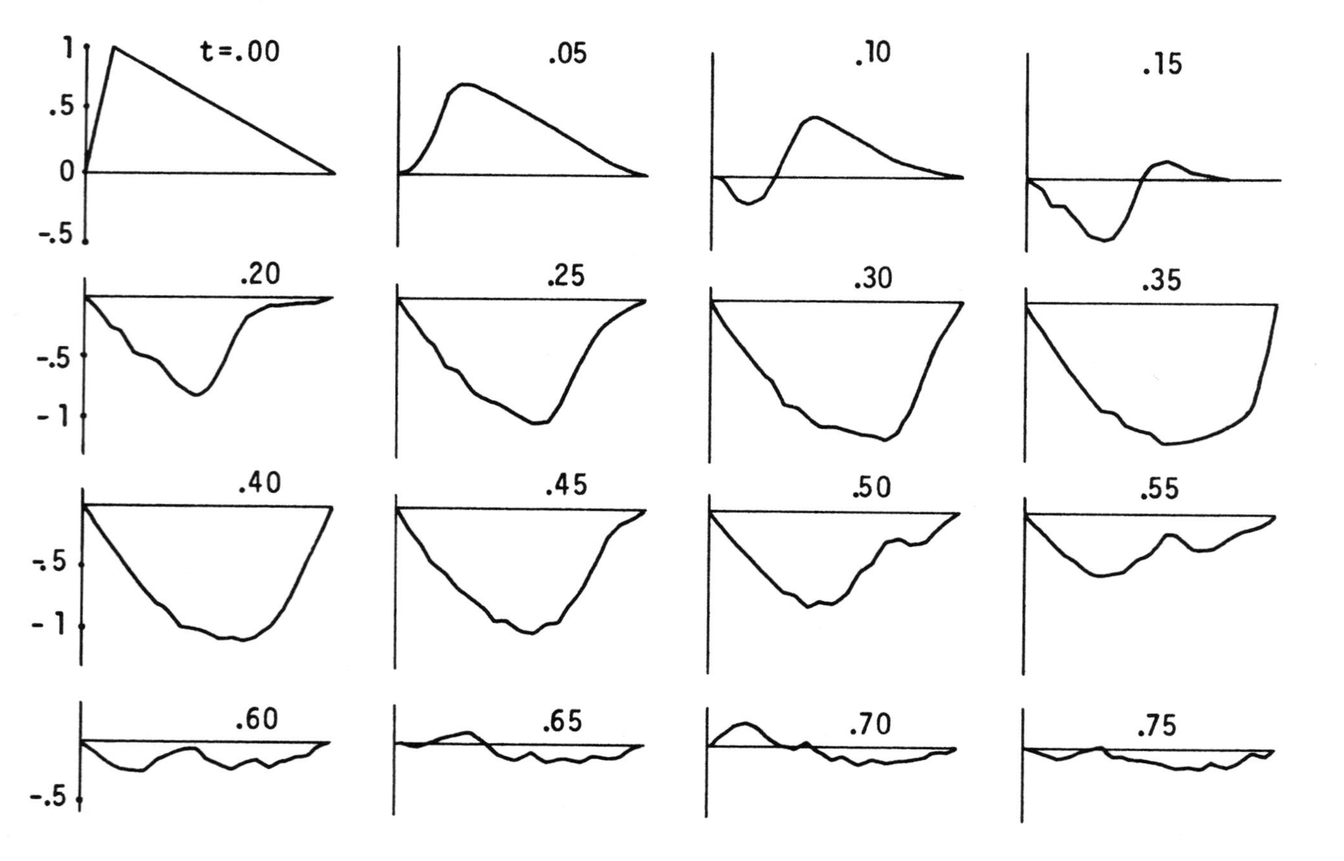

FIGURE 3.6

was obtained as follows. Substitution of the given parameters into (3.5) yields the system

$$
\begin{aligned}
(3.8)\qquad & 12.5\left[1+\left|\frac{y_{j+1}-y_j}{0.1}\right|+0.005\left(\frac{y_{j+1}-y_j}{0.1}\right)^2\right]\times \\
& \qquad \left[\frac{y_{j+1}-y_j}{[(0.01)+(y_{j+1}-y_j)^2]^{1/2}}\right] \\
& -12.5\left[1+\left|\frac{y_j-y_{j-1}}{0.1}\right|+0.005\left(\frac{y_j-y_{j-1}}{0.1}\right)^2\right]\times \\
& \qquad \left[\frac{y_j-y_{j-1}}{[(0.01)+(y_j-y_{j-1})^2]^{1/2}}\right] \\
& = 1.61, \quad j = 1,2,3,\ldots,19\ .
\end{aligned}
$$

System (3.8) was reduced to 10 equations in $y_1, y_2, \ldots, y_{10}$, in which the absolute value signs could be dropped, by assuming the symmetry condition $y_j = y_{20-j}$, $j = 0,1,\ldots,9$, and by assuming the hanging chain condition $y_j > y_{j+1}$, $j = 0, 1,\ldots,9$. The reduced system was solved by the generalized Newton's method and the final answer was checked in (3.8) and verified to be a solution. At the end of six seconds, the string particles were at most 0.0025 from steady state. The total computer time consumed for six seconds of string vibration was under 50 seconds.

Example 2. The string in Example 1 was considered again but with a different initial position. The first moving particle was placed at (0.1, 0.5), the second was placed at (0.2, 1.0), and the remaining were centered on $y = -\frac{5}{9}(x-2)$, as shown at $t = 0.00$ in Figure 3.6. The first 0.75 seconds of motion is shown typically in Figure 3.6. Convergence to the steady S, shown in Figure 3.5, was at a rate comparable to that of Ex-

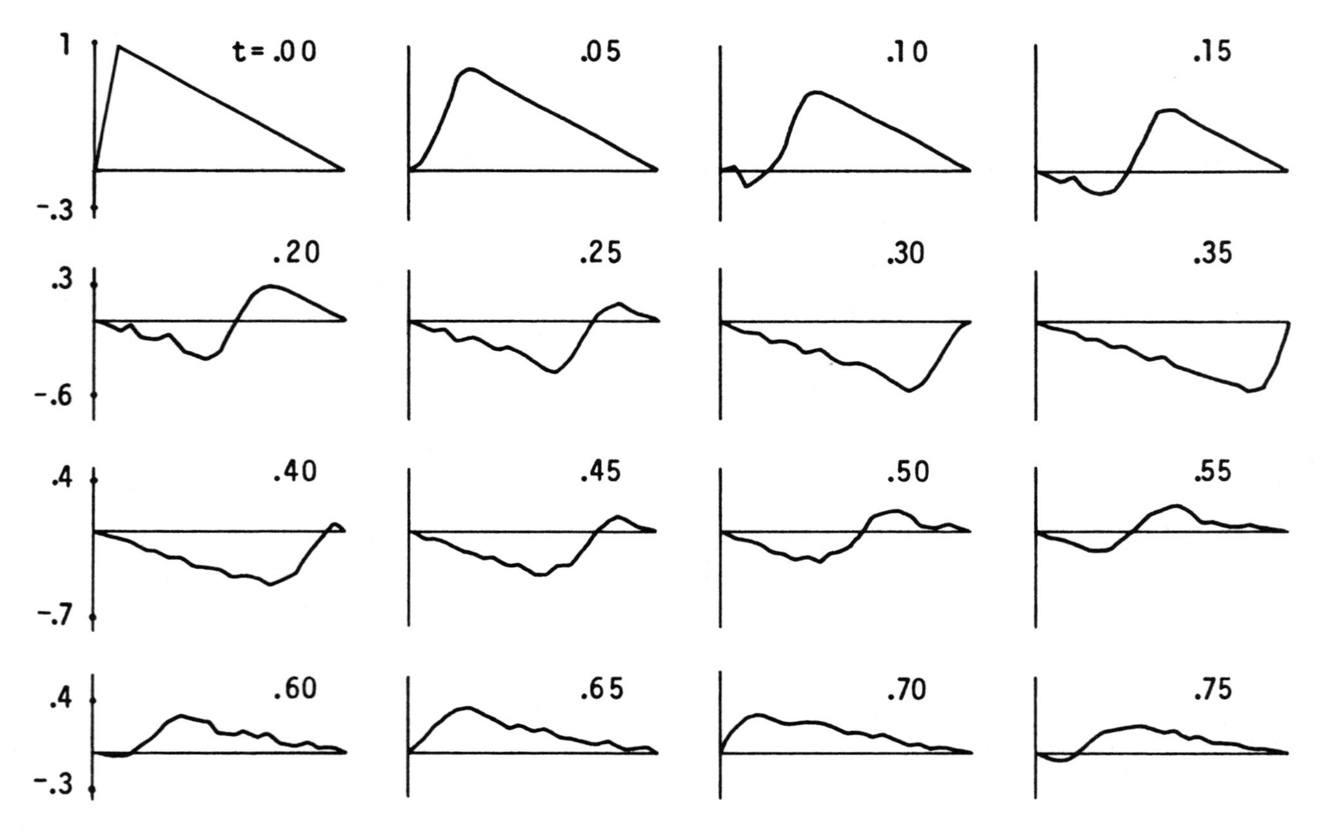

FIGURE 3.7

ample 1.

Example 3. The string of Example 2 was considered again but with zero gravity, that is, with $g = 0$. The first 0.75 seconds of motion are shown typically in Figure 3.7. Convergence to the steady state solution $y_j = 0$, $j = 0,1,2,\ldots,19$, was at a rate comparable to that of Example 2.

Example 4. The string in Example 1 was considered again but with an initial position defined as follows. The fifth moving particle was set at (0.5, -1) and the fifteenth moving particle at (1.5, 1.0). The particles to the left of the fifth were positioned on $y = -2x$, those between the fifth and fifteenth on $y = 2x - 2$, and those to the right of the fifteenth on $y = -2x + 4$. The resulting initial configuration is that shown for $t = 0.00$ in Figure 3.8, where the first 0.95 seconds of motion is shown. Convergence to steady state S, which is shown in Figure 3.5, was at a rate comparable to that of Example 1.

3.5 REMARKS

The intuition used in constructing the examples of Section 3.4 and in determining viable sets of parameter choices can be outlined as follows. A variety of initial conditions and parameters are inserted into (3.2)-(3.4) and the computer is programmed to give 5-10 seconds of vibrations. If no case appears to be approaching an equilibrium position, then Δt and ε are decreased, while α and $\bar{m}$ are increased. When a physically reasonable case results, others can be constructed with a steady state closer to the horizontal by decreasing $\bar{m}$ a small amount, while still others with larger oscillations can be constructed by decreasing α a

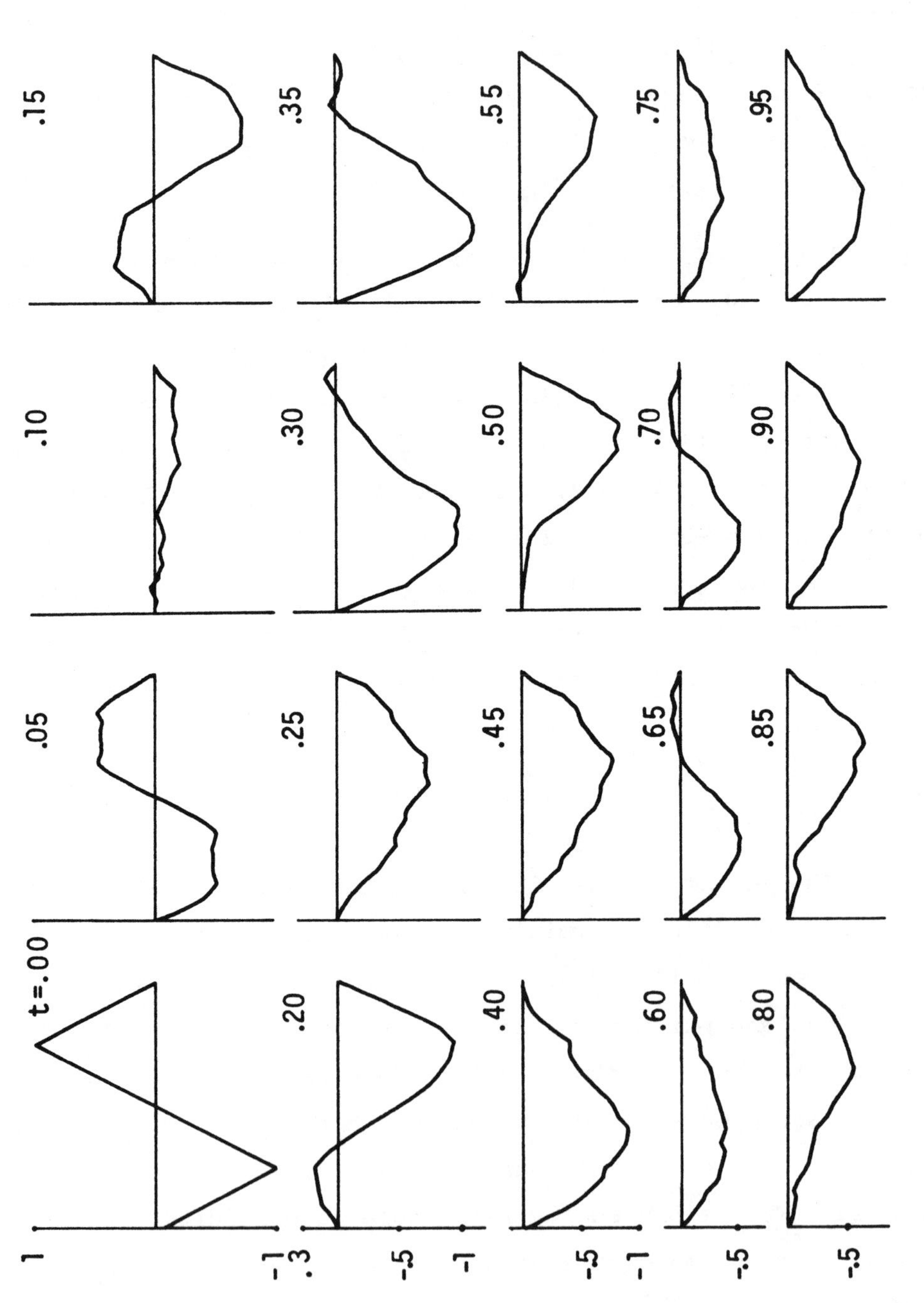
t=.00
.05
.10
.15
.20
.25
.30
.35
.40
.45
.50
.55
.60
.65
.70
.75
.80
.85
.90
.95
1
-1
.3
-.5
-1
-.5
-1
-.5
-.5

FIGURE 3.8

small amount. If a decrease in α or Δx results in instability, then Δt must also be decreased to retain stability.

Now, the examples of Section 3.4 do suggest that the discrete approach to nonlinear string oscillations is reasonable. At present, however, we will not continue this study because we desire first to establish the broad applicability of the discrete model approach. Thus, later, in Chapter IX, we again will consider strings, but with many more particles and with other formulas for tension. Suffice it to say at present that even in so simple an example as that shown in Figure 3.3, the output exhibits clearly a flutter effect between particles which is not a consequence of any continuous model. This flutter effect also will be discussed in greater detail in Chapter IX.

CHAPTER IV - PLANETARY MOTION AND DISCRETE NEWTONIAN GRAVITATION

4.1 INTRODUCTION

Thus far we have been concerned only with motion in one direction of a single particle and of a system of particles. In this chapter, by studying two dimensional motion, we will, in fact, develop the basic methodology necessary for the study of motion in an arbitrary number of dimensions. The prototype physical problem to which our formulas and theorems will be applied is that of the motion of a planet about a stationary sun. Of course, to do this, we will need to develop, in addition, a discrete theory of Newtonian gravitation.

4.2 BASIC PLANAR DYNAMICAL CONCEPTS

Let us show first how one can extend the basic discrete concepts and results of Chapter I to two dimensions. For this purpose, if $\Delta t > 0$ and $t_k = k\Delta t$, $k = 0,1,2,\ldots,n$, let particle P of mass m be located at $\vec{x}_k = (x_k, y_k)$ at time t_k. If $\vec{v}_k = (v_{k,x}, v_{k,y})$ is the velocity of P at t_k, while $\vec{a}_k = (a_{k,x}, a_{k,y})$ is the acceleration of P at t_k, we will assume, in analogy with (1.5) and (1.6), that

$$(4.1)\qquad \frac{\vec{v}_{k+1} + \vec{v}_k}{2} = \frac{\vec{x}_{k+1} - \vec{x}_k}{\Delta t}, \quad k = 0,1,2,\ldots,n-1,$$

$$(4.2)\qquad \vec{a}_{k,x} = \frac{\vec{v}_{k+1} - \vec{v}_k}{\Delta t}, \qquad k = 0,1,2,\ldots,n-1.$$

To relate force and acceleration at each time t_k, we assume a discrete Newtonian equation

$$(4.3)\qquad \vec{F}_k = m\vec{a}_k ,$$

where

$$(4.4)\qquad \vec{F}_k = (F_{k,x}, F_{k,y}) .$$

The work W done by force $\vec{F}$ on particle P from an initial time t_0 to a terminal time t_n is defined by

$$(4.5)\qquad W = \sum_{k=0}^{n-1} [(x_{k+1}-x_k)F_{k,x} + (y_{k+1}-y_k)F_{k,y}] .$$

From (4.1)-(4.4), then,

$$\sum_{k=0}^{n-1} (x_{k+1}-x_k)F_{k,x} = \sum_{0}^{n-1} (x_{k+1}-x_k)ma_{k,x}$$

$$= m \sum_{0}^{n-1} \left(\frac{x_{k+1}-x_k}{\Delta t}\right)(v_{k+1,x}-v_{k,x})$$

$$= \frac{m}{2} \sum_{0}^{n-1} (v_{k+1,x}+v_{k,x})(v_{k+1,x}-v_{k,x})$$

$$= \frac{m}{2} \sum_{0}^{n-1} (v_{k+1,x}^2-v_{k,x}^2)$$

$$= \frac{m}{2} v_{n,x}^2 - \frac{m}{2} v_{0,x}^2 .$$

Similarly,

$$\sum_{k=0}^{n-1} (y_{k+1}-y_k)F_{k,y} = \frac{m}{2} v_{n,y}^2 - \frac{m}{2} v_{0,y}^2 .$$

Thus, from (4.5),

$$(4.6) \qquad W = \frac{m}{2}(v_{n,x}^2+v_{n,y}^2) - \frac{m}{2}(v_{0,x}^2+v_{0,y}^2) .$$

If one defines K_i, the kinetic energy at t_i, by

$$(4.7) \qquad K_i = \frac{m}{2}(v_{i,x}^2+v_{i,y}^2) ,$$

then (4.6) yields, finally, the classical result

$$(4.8) \qquad W = K_n - K_0 .$$

4.3 PLANETARY MOTION AND DISCRETE GRAVITATION

The prototype orbit problem to be considered can be formulated as follows. Let the sun, whose mass is m_1, be positioned at the origin of the XY coordinate system. Let the position, velocity, and mass m of a planet P be known at time t_0. Then, assuming that the sun's motion is negligible, we must determine the position (x_k,y_k) of P at each t_k, $k = 1,2,3,\ldots,n$, if the only acting force is gravitation, and if gravitation is defined by the discrete formulas

$$(4.9) \qquad \vec{F}_k = (F_{k,x},F_{k,y})$$

$$(4.10) \qquad F_{k,x} = -\frac{Gm_1m_2}{r_kr_{k+1}} \cdot \frac{\frac{x_{k+1}+x_k}{2}}{\frac{r_k+r_{k+1}}{2}} \equiv -\frac{Gm_1m_2(x_{k+1}+x_k)}{r_kr_{k+1}(r_k+r_{k+1})}$$

$$(4.11) \qquad F_{k,y} = -\frac{Gm_1m_2(y_{k+1}+y_k)}{r_kr_{k+1}(r_k+r_{k+1})} ,$$

where G is a Newtonian gravitational constant and

(4.12) $$r_k^2 = x_k^2 + y_k^2 .$$

Intuitively, (4.10)-(4.11) express gravitation as a $\frac{1}{r^2}$ law.

4.4 CONSERVATION OF ENERGY

Let us show first that, as is usually deemed desirable for astrophysical models, the model defined in Section 4.3 is conservative. To do this will require a suitable definition of potential energy, which is arrived at as follows.

Consider, again, formula (4.5) for work. Then (4.10)-(4.12) imply

$$W = \sum_{k=0}^{n-1} \left[(x_{k+1}-x_k)\left(-\frac{Gm_1m_2(x_{k+1}+x_k)}{r_kr_{k+1}(r_k+r_{k+1})}\right) + (y_{k+1}-y_k)\left(-\frac{Gm_1m_2(y_{k+1}+y_k)}{r_kr_{k+1}(r_k+r_{k+1})}\right)\right]$$

$$= -Gm_1m_2 \sum_0^{n-1} \left[\frac{x_{k+1}^2-x_k^2+y_{k+1}^2-y_k^2}{r_kr_{k+1}(r_k+r_{k+1})}\right]$$

$$= -Gm_1m_2 \sum_0^{n-1} \left[\frac{r_{k+1}^2 - r_k^2}{r_kr_{k+1}(r_k+r_{k+1})}\right]$$

$$= -Gm_1m_2 \sum_0^{n-1} \left[\frac{r_{k+1} - r_k}{r_k\ r_{k+1}}\right]$$

$$= -Gm_1m_2 \sum_0^{n-1} \left[\frac{1}{r_k} - \frac{1}{r_{k+1}}\right]$$

$$= \frac{-Gm_1m_2}{r_0} + \frac{Gm_1m_2}{r_n} .$$

If one defines the potential energy V_k at t_k by

(4.13) $$V_k = -\frac{Gm_1m_2}{r_k},$$

then

(4.14) $$W = -V_n + V_0 .$$

Hence, (4.8) and (4.14) imply

(4.15) $$K_n + V_n = K_0 + V_0 ,$$

which is, of course, the law of conservation of energy.

Once again, it is worth noting that a telescopic sum, which in this case is

$$\sum_0^{n-1}\left[\frac{1}{r_k} - \frac{1}{r_{k+1}}\right] ,$$

plays a key role in the determination of the potential function V_k.

4.5 EXAMPLE

To illustrate the ease with which the discrete approach can be implemented on a digital computer, consider the normalized orbit problem (Feynman, Leighton and Sands) in which

(4.16) $$Gm_1 = 1$$

and

(4.17) $$x_0 = 0.50,\ y_0 = 0.00,\ v_{0,x} = 0.00,\ v_{0,y} = 1.63.$$

In the classical continuous formulation, the planet's trajectory is an ellipse with semi-major axis $a = 0.746$ and with period $\tau = 4.04$ seconds.

From (4.3) and (4.9)-(4.12), the discrete equations of motion are

$$a_{k,x} = -\frac{x_{k+1} + x_k}{r_k r_{k+1}(r_k+r_{k+1})}, \quad a_{k,y} = -\frac{y_{k+1} + y_k}{r_k r_{k+1}(r_k+r_{k+1})},$$

or, equivalently,

$$(4.18) \quad x_{k+1} = x_k + \frac{\Delta t}{2}(v_{k+1,x} + v_{k,x})$$

$$(4.19) \quad y_{k+1} = y_k + \frac{\Delta t}{2}(v_{k+1,y} + v_{k,y})$$

$$(4.20) \quad v_{k+1,x} = v_{k,x} - \frac{(x_{k+1} + x_k)\Delta t}{(x_k^2+y_k^2)^{\frac{1}{2}}(x_{k+1}^2+y_{k+1}^2)^{\frac{1}{2}}[(x_k^2+y_k^2)^{\frac{1}{2}}+(x_{k+1}^2+y_{k+1}^2)^{\frac{1}{2}}]}$$

$$(4.21) \quad v_{k+1,y} = v_{k,y} - \frac{(y_{k+1} + y_k)\Delta t}{(x_k^2+y_k^2)^{\frac{1}{2}}(x_{k+1}^2+y_{k+1}^2)^{\frac{1}{2}}[(x_k^2+y_k^2)^{\frac{1}{2}}+(x_{k+1}^2+y_{k+1}^2)^{\frac{1}{2}}]}.$$

A solution of (4.18)-(4.21) for each value of $k = 0, 1,2,\ldots,n$, beginning from initial data (4.17) is found by Newton's method with initial guess $x_{k+1}^{(0)} = x_k$, $y_{k+1}^{(0)} = y_k$, $v_{k+1,x}^{(0)} = v_{k,x}$, $v_{k+1,y}^{(0)} = v_{k,y}$. As a typical example of the calculations done, the orbit was generated for $\Delta t = 0.001$ up to $t_{350000} = 350$. The total computing time was under 5 minutes on the UNIVAC 1108. There were 86+ orbits, the 86th of which is shown in Figure 4.1. For this particular orbit, the period is $\tau = 4.05$ and the average of the absolute values of the x intercepts yields $a = 0.746$.

It is of value to note that the particular ordering

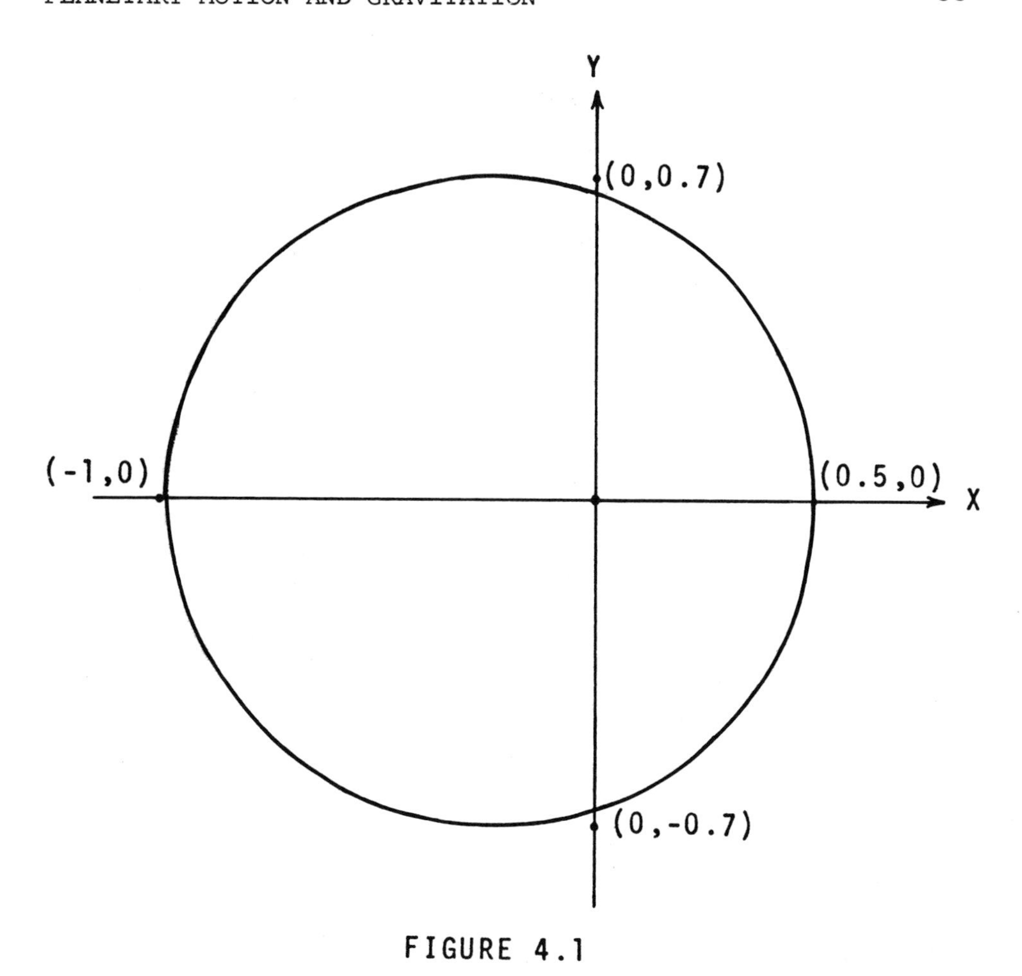

FIGURE 4.1

(4.18)-(4.21) results in very simple formulas for Newtonian iteration.

4.6 REMARKS

Note, first, that the model of Section 4.3 does not suffer from a basic shortcoming of classical gravitation. Newton's formulation was continuous and implied that the force changes magnitude and direction instantaneously if one mass changes position relative to the other. Such a property is considered untenable from the viewpoint of modern physics and has led to other approaches to gravitation, like that of general relativity. The discrete formulation of Section 4.3

implies the acceptable conclusion that the force is known only at each t_k and that its nature is uncertain between these times.

It is also important to note that (4.18)-(4.21) are implicit, thereby requiring the additional work of solving a system of nonlinear algebraic equations at each time step. Thus, though we have gained an energy conserving formulation of gravitational interaction by the particular choices (4.10) and (4.11), we have lost the simplicity of the existence and uniqueness proof of Theorem 1.1. Without such assurance, a priori, it is incumbent upon one to check his numerical solution, a posteriori, at each time step, and, to be completely rigorous, to say that he has generated <u>a</u>, rather than <u>the</u>, solution of a gravitational initial value problem.

Next, it is worth noting that we can extend the force law given in (4.9)-(4.12) to include repulsion. This can be done by rewriting (4.10) and (4.11) as

$$(4.22)\quad F_{k,x} = -\frac{Gm_1m_2(x_{k+1}+x_k)}{r_kr_{k+1}(r_k+r_{k+1})} + \frac{Hm_1m_2\left[\sum_{j=0}^{m-2}(r_k^j r_{k+1}^{m-j-2})\right](x_{k+1}+x_k)}{r_k^{m-1}r_{k+1}^{m-1}(r_{k+1}+r_k)}$$

$$(4.23)\quad F_{k,y} = -\frac{Gm_1m_2(y_{k+1}+y_k)}{r_kr_{k+1}(r_k+r_{k+1})} + \frac{Hm_1m_2\left[\sum_{j=0}^{m-2}(r_k^j r_{k+1}^{m-j-2})\right](y_{k+1}+y_k)}{r_k^{m-1}r_{k+1}^{m-1}(r_{k+1}+r_k)}$$

where H is a parameter of repulsion and $m \ge 2$. Intuitively, (4.22)-(4.23) express attraction as a $\frac{1}{r^2}$ component and repulsion as a $\frac{1}{r^m}$ component, $m \ge 2$.

To show that energy is conserved under (4.22)-(4.23), we need only establish a new potential function V_k and repeat the argument that led to (4.15). Since we know by (4.15) that the attracting portion of (4.22) is energy conserving, let us simply show that the same is true of the repelling portion, which will establish conservation. Hence, (4.5), (4.22) and (4.23) imply

$$W = \sum_{k=0}^{n-1} \left\{ (x_{k+1}-x_k) \frac{Hm_1m_2\left[\sum_{j=0}^{m-2} (r_k^j r_{k+1}^{m-j-2})\right](x_{k+1}+x_k)}{r_k^{m-1} r_{k+1}^{m-1} (r_{k+1}+r_k)} \right.$$

$$\left. + (y_{k+1}-y_k) \frac{Hm_1m_2\left[\sum_{j=0}^{m-2} (r_k^j r_{k+1}^{m-j-2})\right](y_{k+1}+y_k)}{r_k^{m-1} r_{k+1}^{m-1} (r_{k+1}+r_k)} \right\}$$

$$= Hm_1m_2 \sum_{k=0}^{n-1} \left\{ \frac{\left[\sum_{j=0}^{m-2} (r_k^j r_{k+1}^{m-j-2})\right]}{r_k^{m-1} r_{k+1}^{m-1} (r_{k+1}+r_k)} [x_{k+1}^2 - x_k^2 + y_{k+1}^2 - y_k^2] \right\}$$

$$= Hm_1m_2 \sum_{k=0}^{n-1} \left\{ \frac{\left[\sum_{j=0}^{m-2} (r_k^j r_{k+1}^{m-j-2})\right] (r_{k+1}-r_k)}{r_k^{m-1} r_{k+1}^{m-1}} \right\}.$$

But,

$$\left[\sum_{j=0}^{m-2} (r_k^j r_{k+1}^{m-j-2})\right](r_{k+1}-r_k) \equiv r_{k+1}^{m-1} - r_k^{m-1} ,$$

so that

$$W = Hm_1m_2 \sum_{k=0}^{n-1} \left(\frac{r_{k+1}^{m-1} - r_k^{m-1}}{r_k^{m-1} r_{k+1}^{m-1}} \right) = Hm_1m_2 \sum_{k=0}^{n-1} \left(\frac{1}{r_k^{m-1}} - \frac{1}{r_{k+1}^{m-1}} \right)$$

$$= Hm_1m_2 \left(\frac{1}{r_0^{m-1}} - \frac{1}{r_n^{m-1}} \right) .$$

Defining the potential function V_k by

$$V_k = \frac{Hm_1m_2}{r_k^{m-1}}$$

yields

$$(4.24)\quad W = V_0 - V_n ,$$

and conservation follows from (4.8) and (4.24).

Finally, the latter discussion implies that (4.22) and (4.23) can be extended in an energy conserving fashion to include attraction as a $1/(r^n)$ component, $n \geq 2$, and repulsion as a $1/(r^m)$ component, $m \geq 2$, if the force is defined by

$$(4.24)\quad F_{k,x} = -\frac{Gm_1m_2\left[\sum_{j=0}^{n-2}(r_k^j r_{k+1}^{n-j-2})\right](x_{k+1}+x_k)}{r_k^{n-1}r_{k+1}^{n-1}(r_{k+1}+r_k)} + \frac{Hm_1m_2\left[\sum_{j=0}^{m-2}(r_k^j r_{k+1}^{m-j-2})\right](x_{k+1}+x_k)}{r_k^{m-1}r_{k+1}^{m-1}(r_{k+1}+r_k)}$$

$$(4.25)\quad F_{k,y} = -\frac{Gm_1m_2\left[\sum_{j=0}^{n-2}(r_k^j r_{k+1}^{n-j-2})\right](y_{k+1}+y_k)}{r_k^{n-1}r_{k+1}^{n-1}(r_{k+1}+r_k)} + \frac{Hm_1m_2\left[\sum_{j=0}^{m-2}(r_k^j r_{k+1}^{m-j-2})\right](y_{k+1}+y_k)}{r_k^{m-1}r_{k+1}^{m-1}(r_{k+1}+r_k)} .$$

Forces defined by (4.24)-(4.25) are fundamental in the classical approach to molecular dynamics (Feynman *et al.*).

CHAPTER V - THE THREE-BODY PROBLEM

5.1 INTRODUCTION

Since the preceding chapter was devoted to the motion of a single particle in the plane, it is natural that we now study the motion of a system of particles in the plane. The simplest such system, for which no general classical analysis exists, is that of three nondegenerate particles in motion under the influence only of gravitation. We will show next how to solve this problem from the discrete point of view.

5.2 THE EQUATIONS OF MOTION

For $\Delta t > 0$ and $t_k = k\Delta t$, $k = 0,1,2,\ldots,n$, and for each of $i = 1,2,3$, let particle P_i of mass m_i be located at $\vec{x}_{i,k} = (x_{i,k}, y_{i,k})$, have velocity $\vec{v}_{i,k} = (v_{i,k,x}, v_{i,k,y})$, and acceleration $\vec{a}_{i,k} = (a_{i,k,x}, a_{i,k,y})$ at time t_k. In analogy with (1.5), (1.6), (4.1) and (4.2), let

$$(5.1) \qquad \frac{\vec{v}_{i,k+1} + \vec{v}_{i,k}}{2} = \frac{\vec{x}_{i,k+1} - \vec{x}_{i,k}}{\Delta t},$$

$$i = 1,2,3; \quad k = 0,1,2,\ldots,n-1,$$

(5.2) $$\vec{a}_{i,k} = \frac{\vec{v}_{i,k+1} - \vec{v}_{i,k}}{\Delta t},$$

$$i = 1,2,3; \quad k = 0,1,2,\ldots,n-1 .$$

Of course, (5.1)-(5.2) differ from (4.1)-(4.2) only by the addition of the subscript i, which enables one to associate a given velocity and acceleration with a particular particle in the system.

To relate force and acceleration, we assume a discrete Newtonian equation

(5.3) $$\vec{F}_{i,k} = m_i \vec{a}_{i,k}; \quad i = 1,2,3; \quad k = 0,1,2,\ldots,n-1 ,$$

where

(5.4) $$\vec{F}_{i,k} = (F_{i,k,x}, F_{i,k,y}) .$$

The work W_i done by $\vec{F}_{i,k}$ on P_i from initial time t_0 to terminal time t_n is defined by

(5.5) $$W_i = \sum_{k=0}^{n-1} [(x_{i,k+1}-x_{i,k})F_{i,k,x}+(y_{i,k+1}-y_{i,k})F_{i,k,y}],$$

while the total work W is defined by

(5.6) $$W = \sum_{i=1}^{3} W_i .$$

The exact derivation which yielded (4.6) implies, again,

(5.7) $$W_i = \frac{m_i}{2}(v_{i,n,x}^2+v_{i,n,y}^2) - \frac{m_i}{2}(v_{i,0,x}^2+v_{i,0,y}^2) ,$$

so that if the kinetic energy $K_{i,k}$ of P_i at t_k is defined by

(5.8) $$K_{i,k} = \frac{m_i}{2}(v_{i,k,x}^2+v_{i,k,y}^2) ,$$

then

(5.9) $$W_i = K_{i,n} - K_{i,0} .$$

Defining the kinetic energy K_k of the system at time t_k by

(5.10) $$K_k = \sum_{i=1}^{3} K_{i,k}$$

yields, finally,

(5.11) $$W = K_n - K_0 .$$

Next, the precise structure of the force components of (5.4) is given as follows. If $r_{ij,k}$ is the distance between P_i and P_j at time t_k, then, in analogy with (4.10) and (4.11), set

(5.12) $$F_{1,k,x} = -\frac{Gm_1m_2[(x_{1,k+1}+x_{1,k})-(x_{2,k+1}+x_{2,k})]}{r_{12,k}r_{12,k+1}(r_{12,k}+r_{12,k+1})} - \frac{Gm_1m_3[(x_{1,k+1}+x_{1,k})-(x_{3,k+1}+x_{3,k})]}{r_{13,k}r_{13,k+1}(r_{13,k}+r_{13,k+1})}$$

(5.13) $$F_{2,k,x} = -\frac{Gm_1m_2[(x_{2,k+1}+x_{2,k})-(x_{1,k+1}+x_{1,k})]}{r_{12,k}r_{12,k+1}(r_{12,k}+r_{12,k+1})} - \frac{Gm_2m_3[(x_{2,k+1}+x_{2,k})-(x_{3,k+1}+x_{3,k})]}{r_{23,k}r_{23,k+1}(r_{23,k}+r_{23,k+1})}$$

(5.14) $$F_{3,k,x} = -\frac{Gm_1m_3[(x_{3,k+1}+x_{3,k})-(x_{1,k+1}+x_{1,k})]}{r_{13,k}r_{13,k+1}(r_{13,k}+r_{13,k+1})} - \frac{Gm_2m_3[(x_{3,k+1}+x_{3,k})-(x_{2,k+1}+x_{2,k})]}{r_{23,k+1}r_{23,k}(r_{23,k}+r_{23,k+1})} ,$$

while $F_{1,k,y}$, $F_{2,k,y}$, $F_{3,k,y}$ are defined by substituting y for x in (5.12), (5.13) and (5.14), respectively.

5.3 CONSERVATION OF ENERGY

To establish the conservation of energy, consider, again, (5.6). Substitution of (5.12)-(5.14) and the corresponding formulas for $F_{1,k,y}$, $F_{2,k,y}$ and $F_{3,k,y}$ into (5.6) yields readily

$$W = -Gm_1m_2 \sum_{k=0}^{n-1} \left(\frac{r_{12,k+1}-r_{12,k}}{r_{12,k}\, r_{12,k+1}}\right)$$

$$- Gm_1m_3 \sum_{k=0}^{n-1} \left(\frac{r_{13,k+1}-r_{13,k}}{r_{13,k}\, r_{13,k+1}}\right)$$

$$- Gm_2m_3 \sum_{k=0}^{n-1} \left(\frac{r_{23,k+1}-r_{23,k}}{r_{23,k}\, r_{23,k+1}}\right)$$

$$= -Gm_1m_2 \left(\frac{1}{r_{12,0}} - \frac{1}{r_{12,n}}\right) - Gm_1m_3 \left(\frac{1}{r_{13,0}} - \frac{1}{r_{13,n}}\right)$$

$$- Gm_2m_3 \left(\frac{1}{r_{23,0}} - \frac{1}{r_{23,n}}\right) .$$

Defining the potential energy $V_{ij,k}$ of the pair P_i and P_j at t_k by

$$V_{ij,k} = -G\,\frac{m_i m_j}{r_{ij,k}}$$

implies then that

$$(5.15) \qquad W = V_{12,0} + V_{13,0} + V_{23,0} - V_{12,n} - V_{13,n} - V_{23,n}.$$

If the potential energy V_k of the system at time t_k is defined by

$$V_k = V_{12,k} + V_{13,k} + V_{23,k} ,$$

then (5.15) implies

$$(5.16) \quad W = V_0 - V_n .$$

Finally, elimination of W between (5.11) and (5.16) yields the desired result

$$K_n + V_n = K_0 + V_0 .$$

5.4 SOLUTION OF THE DISCRETE THREE-BODY PROBLEM

Though we can now describe a general algorithm for generating a solution of the three-body problem, it is more instructive to consider a particular problem and to show, in detail, how to solve it. The reasoning required for other problems is entirely analogous. Consider, therefore, as shown in Figure 5.1, three particles P_1, P_2, P_3, of equal

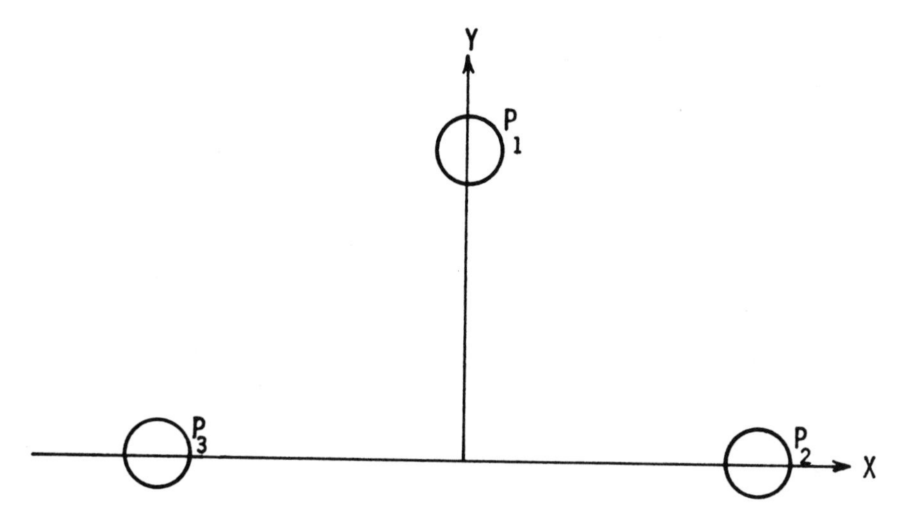

FIGURE 5.1

masses which are normalized so that

(5.17) $m_1 = m_2 = m_3 = 10, \quad G = 1$.

Let the initial positions and velocities be given by $x_{1,0} = 0$, $y_{1,0} = 100$, $x_{2,0} = 100$, $y_{2,0} = 0$, $x_{3,0} = -100$, $y_{3,0} = 0$, $v_{1,0,x} = 0$, $v_{1,0,y} = -10$, $v_{2,0,x} = -10$, $v_{2,0,y} = 0$, $v_{3,0,x} = 9.9$, $v_{3,0,y} = 0$. From (5.1)-(5.4), one can, as in (4.18)-(4.21), rewrite the equations of motion as follows:

(5.18) $$x_{i,k+1} = x_{i,k} + \frac{\Delta t}{2}(v_{i,k+1,x}+v_{i,k,x}), \quad i = 1,2,3$$

(5.19) $$y_{i,k+1} = y_{i,k} + \frac{\Delta t}{2}(v_{i,k+1,y}+v_{i,k,y}), \quad i = 1,2,3$$

(5.20) $$v_{1,k+1,x} = v_{1,k,x} - 10\Delta t\left\{\frac{(x_{1,k+1}+x_{1,k})-(x_{2,k+1}+x_{2,k})}{r_{12,k}r_{12,k+1}(r_{12,k}+r_{12,k+1})} + \frac{(x_{1,k+1}+x_{1,k})-(x_{3,k+1}+x_{3,k})}{r_{13,k}r_{13,k+1}(r_{13,k}+r_{13,k+1})}\right\}$$

(5.21) $$v_{1,k+1,y} = v_{1,k,y} - 10\Delta t\left\{\frac{(y_{1,k+1}+y_{1,k})-(y_{2,k+1}+y_{2,k})}{r_{12,k}r_{12,k+1}(r_{12,k}+r_{12,k+1})} + \frac{(y_{1,k+1}+y_{1,k})-(y_{3,k+1}+y_{3,k})}{r_{13,k}r_{13,k+1}(r_{13,k}+r_{13,k+1})}\right\}$$

$$(5.22)\quad v_{2,k+1,x} = v_{2,k,x} - 10\Delta t \left\{ \frac{(x_{2,k+1}+x_{2,k})-(x_{1,k+1}+x_{1,k})}{r_{12,k}r_{12,k+1}(r_{12,k}+r_{12,k+1})} + \frac{(x_{2,k+1}+x_{2,k})-(x_{3,k+1}+x_{3,k})}{r_{23,k}r_{23,k+1}(r_{23,k}+r_{23,k+1})} \right\}$$

$$(5.23)\quad v_{2,k+1,y} = v_{2,k,y} - 10\Delta t \left\{ \frac{(y_{2,k+1}+y_{2,k})-(y_{1,k+1}+y_{1,k})}{r_{12,k}r_{12,k+1}(r_{12,k}+r_{12,k+1})} + \frac{(y_{2,k+1}+y_{2,k})-(y_{3,k+1}+y_{3,k})}{r_{23,k}r_{23,k+1}(r_{23,k}+r_{23,k+1})} \right\}$$

$$(5.24)\quad v_{3,k+1,x} = v_{3,k,x} - 10\Delta t \left\{ \frac{(x_{3,k+1}+x_{3,k})-(x_{1,k+1}+x_{1,k})}{r_{13,k}r_{13,k+1}(r_{13,k}+r_{13,k+1})} + \frac{(x_{3,k+1}+x_{3,k})-(x_{2,k+1}+x_{2,k})}{r_{23,k+1}r_{23,k}(r_{23,k}+r_{23,k+1})} \right\}$$

$$(5.25)\quad v_{3,k+1,y} = v_{3,k,y} - 10\Delta t \left\{ \frac{(y_{3,k+1}+y_{3,k})-(y_{1,k+1}+y_{1,k})}{r_{13,k}r_{13,k+1}(r_{13,k}+r_{13,k+1})} + \frac{(y_{3,k+1}+y_{3,k})-(y_{2,k+1}+y_{2,k})}{r_{23,k+1}r_{23,k}(r_{23,k}+r_{23,k+1})} \right\}$$

and

$$(5.26)\quad r_{ij,k} = [(x_{i,k}-x_{j,k})^2 + (y_{i,k}-y_{j,k})^2]^{\frac{1}{2}} .$$

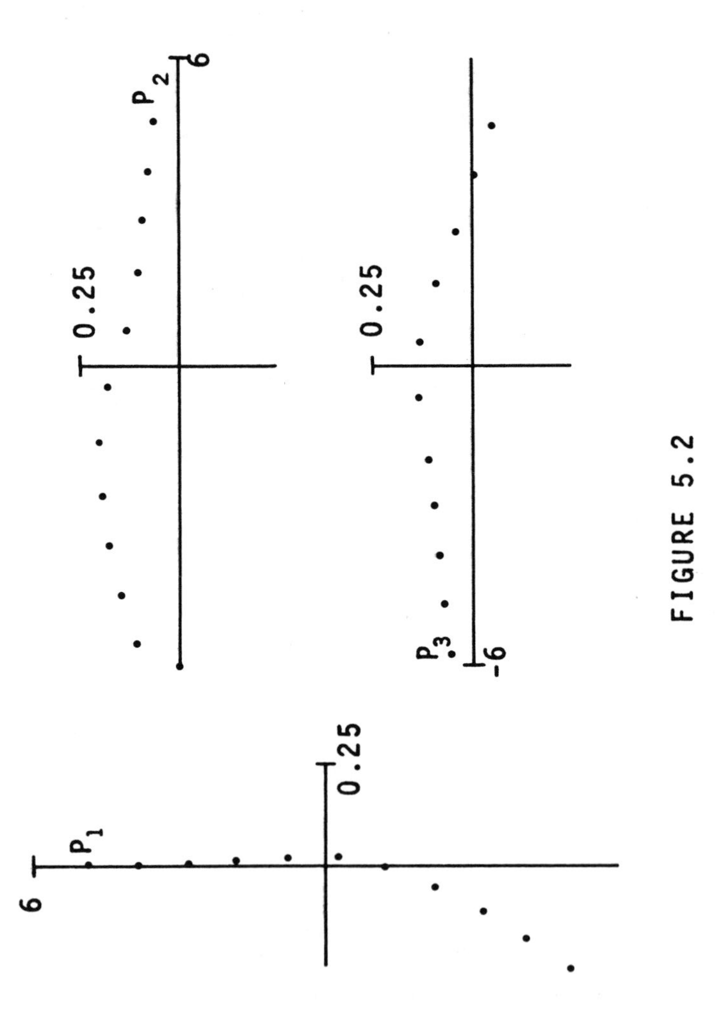

FIGURE 5.2

The solution of the twelve equations (5.18)-(5.25) for the twelve unknowns $x_{i,k+1}$, $y_{i,k+1}$, $v_{i,k+1,x}$, $v_{i,k+1,y}$, $i=1,2,3$, for each value of $k = 0,1,2,\ldots,n-1$, from the initial data is found by Newton's method with initial guess $x^{(0)}_{i,k+1} = x_{i,k}$, $y^{(0)}_{i,k+1} = y_{i,k}$, $v^{(0)}_{i,k+1,x} = v_{i,k,x}$, $v^{(0)}_{i,k+1,y} = v_{i,k,y}$. In Figure 5.2 are shown for $\Delta t = 0.1$ the deflections of the particles from times t_{95} to t_{105}. The motion of each particle is shown separately and the labels P_i, $i = 1,2,3$ are affixed at their positions corresponding to t_{95}. The running time for one thousand time steps was under twenty seconds.

5.5 THE OSCILLATORY NATURE OF PLANETARY PERIHELION MOTION

The methodology developed in Section 5.4 allows us to examine more refined aspects of planetary motion than those of Chapter IV. Let us therefore explore the nature of perihelion motion, and, for this purpose, let us begin with examples in which perihelion motion is large. In each example, the time step is $\Delta t = 0.001$ and CGS units are used, so that $G = (6.67)10^{-8}$. Verification of orbital motion was checked by direct substitution into Newton's version of Kepler's third law, that is,

$$(5.27) \qquad (m_1 + m_2)\tau^2 = \frac{4\pi^2}{G} a^3 ,$$

where τ is the period and a is half the length of the major axis.

<u>Example 1.</u> Consider the three-body problem for particles P_1, P_2, and P_3 with the following initial data:

$$m_1 = (6.67)^{-1}10^8 , \quad m_2 = (6.67)^{-1}10^6 , \quad m_3 = (6.67)^{-1}10^5$$

$$x_{1,0} = 0 \qquad x_{2,0} = 0.5 \qquad x_{3,0} = -1$$

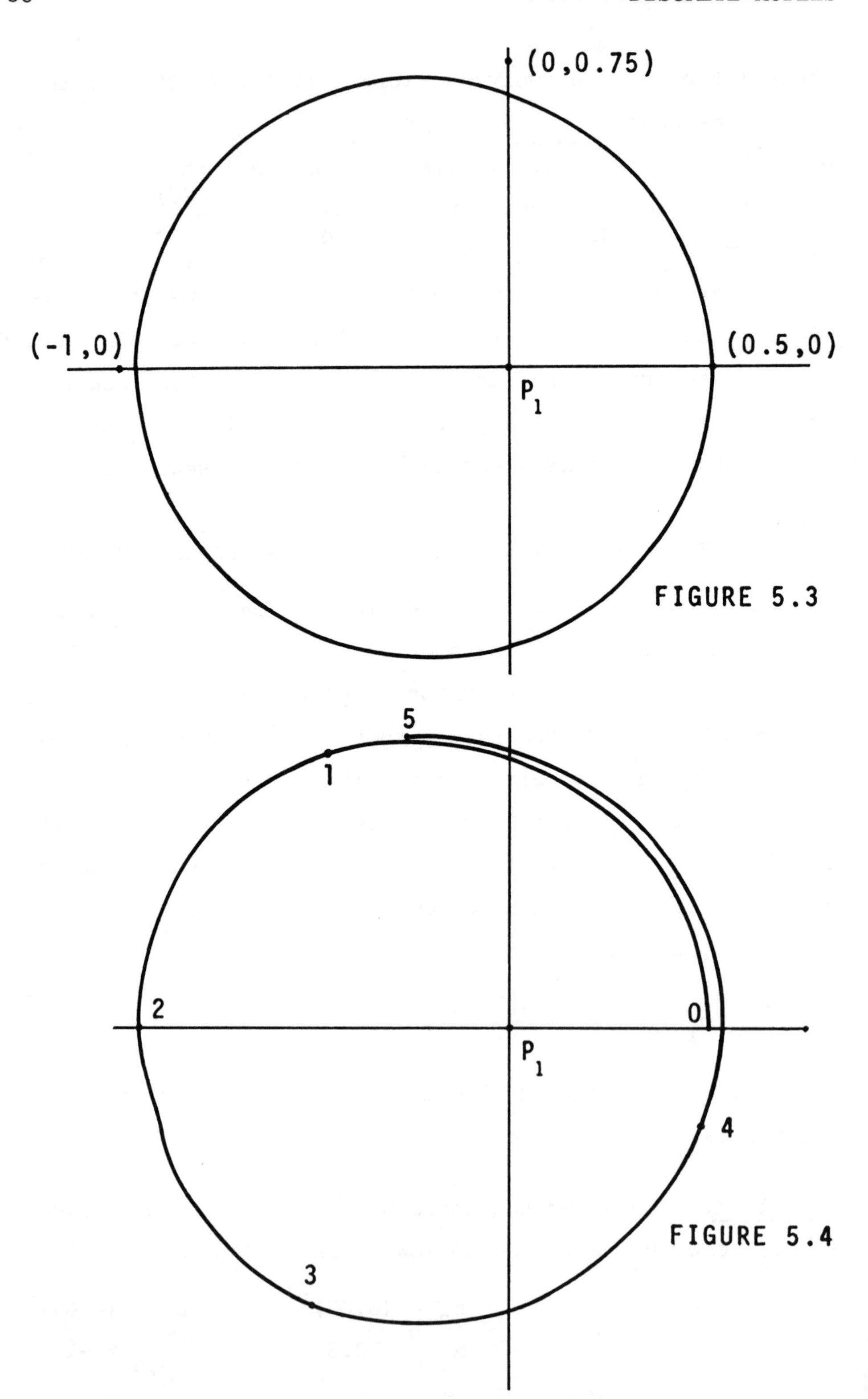

FIGURE 5.3

FIGURE 5.4

$$y_{1,0} = 0 \qquad y_{2,0} = 0 \qquad y_{3,0} = 8$$
$$v_{1,0,x} = 0 \qquad v_{2,0,x} = 0 \qquad v_{3,0,x} = 0$$
$$v_{1,0,y} = 0 \qquad v_{2,0,y} = 1.63 \qquad v_{3,0,y} = -3.75.$$

In the absence of P_3, the motion of P_2 relative to P_1 is the orbit shown in Figure 5.3, for which the period is $\tau = 3.901$ and half the length of the major axis is $a = 0.730$. The initial data for P_3 was chosen so that this particle begins its motion at a relatively large distance from both P_1 and P_2, arrives in the vicinity of (-1,0) almost simultaneously with P_2, and proceeds past (-1,0) at a relatively high velocity, assuring only a short period of strong gravitational attraction. Particles P_2 and P_3 come closest in the third quadrant at t_{2125}, when P_2 is at (-0.9296,-0.1108) and P_3 is at (-0.9325,-0.1012). The effect of the interaction is to deflect P_2 outward, as is seen clearly in Figure 5.4, where the motion of P_2, relative to P_1, has been plotted from t_0 to t_{5000}, with the integer labels $n = 0,1,2,3,4,5$, marking the positions t_{1000n}. After having been deflected, P_2 goes into the new orbit about P_1 which is shown in Figure 5.5. The end points of the new major axis are (0.4943,0.1664) and (-0.9105,-0.3075), so that $a = 0.74135$. The new period is $\tau = 3.9905$. The perihelion motion is measured by the angle of inclination θ of the new major axis with the X axis, and is given by $\tan\theta = 0.34$. Note that the perihelion motion of this example is positive.

Example 2. The data of Example 1 were changed only for P_3 by setting $x_{3,0} = -0.5$, $y_{3,0} = 8.0$, $v_{3,0,x} = -0.25$, $v_{3,0,y} = -4.00$. This time the strongest gravitational effect between P_2 and P_3 occurs in the second quadrant at t_{1966} when P_2 is at (-0.94582,0.01950) and P_3 is at (-0.94418,0.01796),

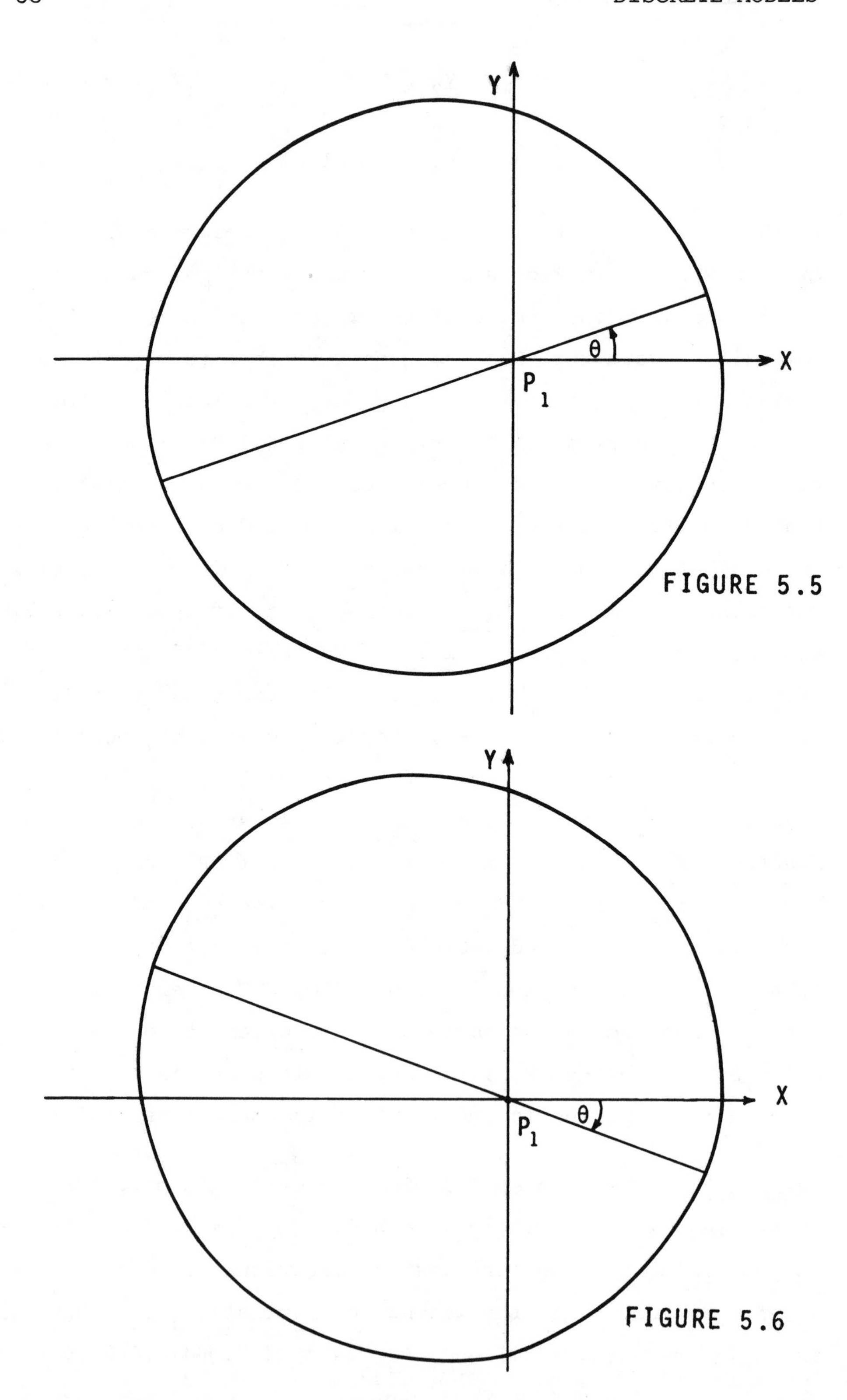

FIGURE 5.5

FIGURE 5.6

and P_2 is perturbed into the new orbit shown in Figure 5.6. The end points of the new major axis are (0.50724, -0.18349) and (-0.92692, 0.33474), so that $a = 0.76246$, and the new period is $\tau = 4.162$. The resulting perihelion motion is now negative, since the angle θ of the new major axis with the X axis is given by $\tan \theta = -0.36$.

From the above and similar examples, it follows that the major axis of P_2 is deflected in the same direction as is P_2. In actual planetary motions, as, for example, in a Sun-Mercury-Venus system, it can be concluded that when Mercury and Venus are relatively close in the first or in the third quadrants, the perihelion of Mercury must be perturbed a small amount in the positive angular direction, while relative closeness in the second or in the fourth quadrants must result in a small negative angular perturbation. All such possibilities can occur for the motions of Mercury and Venus and, because of the differences in their orbital eccentricities and the incommensurability of the orbital periods of these planets, the perihelion motion of Mercury is a complex, nonlinear, oscillatory motion. These conclusions were verified on the computer with ten full orbits of Mercury.

5.6 REMARKS

Note that the motion of the center of gravity of a three-body system is, as in classical mechanics, one dimensional. This can be shown as follows.

Let $t_k = k\Delta t$, $k = 0,1,2,\ldots,n$. From (5.3) and (5.12)-(5.13), one has

$$m_1 a_{1,k,x} + m_2 a_{2,k,x} + m_3 a_{3,k,x} = 0, \quad k \geq 0. \tag{5.28}$$

Hence,

$$(5.29)\quad m_1(v_{1,k+1,x} - v_{1,k,x}) + m_2(v_{2,k+1,x} - v_{2,k,x}) + m_3(v_{3,k+1,x} - v_{3,k,x}) = 0 .$$

Summing both sides of (5.29) over k from 0 to $j-1$, where $j\geq 1$, yields

$$(5.30)\quad m_1(v_{1,j,x} - v_{1,0,x}) + m_2(v_{2,j,x} - v_{2,0,x}) + m_3(v_{3,j,x} - v_{3,0,x}) = 0 .$$

However, since (5.30) is valid even if $j = 0$, it follows that

$$(5.31)\quad m_1 v_{1,j,x} + m_2 v_{2,j,x} + m_3 v_{3,j,x} = c_1, \quad j\geq 0 ,$$

where

$$(5.32)\quad c_1 = m_1 v_{1,0,x} + m_2 v_{2,0,x} + m_3 v_{3,0,x} .$$

Since (5.31) is valid for any j, it must be valid if j is replaced by $j + 1$, so that

$$(5.33)\quad m_1 v_{1,j+1,x} + m_2 v_{2,j+1,x} + m_3 v_{3,j+1,x} = c_1 .$$

Addition of (5.31) and (5.33) then yields

$$m_1\left(\frac{v_{1,j+1,x}+v_{1,j,x}}{2}\right) + m_2\left(\frac{v_{2,j+1,x}+v_{2,j,x}}{2}\right) + m_3\left(\frac{v_{3,j+1,x}+v_{3,j,x}}{2}\right) = c_1 ,$$

or, equivalently,

$$(5.34)\quad m_1(x_{1,j+1}-x_{1,j}) + m_2(x_{2,j+1}-x_{2,j}) + m_3(x_{3,j+1}-x_{3,j}) = c_1\Delta t, \quad j\geq 0.$$

Summing both sides of (5.34) with respect to j from 0 to $n-1$, for $n\geq 1$, yields

$$(5.35)\quad m_1(x_{1,n}-x_{1,0}) + m_2(x_{2,n}-x_{2,0}) + m_3(x_{3,n}-x_{3,0}) = c_1t_n.$$

However, (5.35) is valid also for $n = 0$, so that

$$(5.36)\quad m_1x_{1,n} + m_2x_{2,n} + m_3x_{3,n} = c_1t_n + c_2,\quad n\geq 0,$$

where

$$(5.37)\quad c_2 = m_1x_{1,0} + m_2x_{2,0} + m_3x_{3,0}\ .$$

In a fashion analogous to the derivation of (5.36), it follows also that

$$(5.38)\quad m_1y_{1,n} + m_2y_{2,n} + m_3y_{3,n} = d_1t_n + d_2,\ n\geq 0,$$

where

$$d_1 = m_1v_{1,0,y} + m_2v_{2,0,y} + m_3v_{3,0,y}$$

$$d_2 = m_1y_{1,0} + m_2y_{2,0} + m_3y_{3,0}\ .$$

Finally, if one sets

$$M = m_1 + m_2 + m_3$$

and lets $(\bar{x}_n,\bar{y}_n)$ be the center of gravity of the system at time t_n, then (5.36) and (5.37) imply

$$M\bar{x}_n = c_1t_n + c_2,\quad n\geq 0$$

$$M\bar{y}_n = d_1t_n + d_2,\quad n\geq 0\ ,$$

from which it follows that the motion of the center of gravity is lineal.

Note, incidentally, that the conservation of linear momentum follows directly from (5.31) and the corresponding equation for the y component.

CHAPTER VI - THE n-BODY PROBLEM

6.1 INTRODUCTION

In this chapter we will continue in the spirit of the last chapter and formulate an energy conserving, implicit approach to the n-body problem. Our formulation will be sufficiently general to include both classical molecular and Newtonian interaction. Nonlinear models of the important phenomena of heat transfer and elastic bending will be developed and applied to discrete rods, or bars.

6.2 DISCRETE n-BODY INTERACTION

Again, for positive time step Δt, let $t_k = k\Delta t$, $k = 0,1,2,\ldots$. At time t_k let particle P_i of mass m_i be located at $\vec{x}_{i,k} = (x_{i,k}, y_{i,k})$, have velocity $\vec{v}_{i,k} = (v_{i,k,x}, v_{i,k,y})$, and have acceleration $\vec{a}_{i,k} = (a_{i,k,x}, a_{i,k,y})$, for $i = 1,2,\ldots,n$. Position, velocity, and acceleration are assumed to be related by

$$(6.1)\qquad \frac{\vec{v}_{i,k+1} + \vec{v}_{i,k}}{2} = \frac{\vec{x}_{i,k+1} - \vec{x}_{i,k}}{\Delta t}$$

$$(6.2)\qquad \vec{a}_{i,k} = \frac{\vec{v}_{i,k+1} - \vec{v}_{i,k}}{\Delta t} .$$

If $\vec{F}_{i,k} = (F_{i,k,x}, F_{i,k,y})$ is the force acting on P_i at time t_k, then force and acceleration are assumed to be related by the discrete dynamical equation

$$(6.3) \qquad \vec{F}_{i,k} = m_i \vec{a}_{i,k} \, .$$

In particular, we now choose $\vec{F}_{i,k}$ to have a component of attraction which behaves like $\frac{G}{r^\alpha}$ and a component of repulsion which behaves like $\frac{H}{r^\beta}$, where G, H, α and β are nonnegative parameters which are fixed in any particular problem with $\alpha \geq 2$, $\beta \geq 2$, and where r is the distance between a given pair of particles. For this purpose, let $r_{ij,k}$ be the distance between P_i and P_j at t_k. Then $\vec{F}_{i,k}$, the force exerted on P_i by the remaining particles, is defined, in analogy with (4.24) and (4.25), by

$$(6.4) \qquad \vec{F}_{i,k} = m_i \sum_{\substack{j=1 \\ j \neq i}}^{n} \left\{ m_j \left(- \frac{G \sum_{\xi=0}^{\alpha-2} (r_{ij,k}^{\xi} \, r_{ij,k+1}^{\alpha-\xi-2})}{r_{ij,k}^{\alpha-1} \, r_{ij,k+1}^{\alpha-1} \, (r_{ij,k+1} + r_{ij,k})} + \frac{H \sum_{\xi=0}^{\beta-2} (r_{ij,k}^{\xi} \, r_{ij,k+1}^{\beta-\xi-2})}{r_{ij,k}^{\beta-1} \, r_{ij,k+1}^{\beta-1} (r_{ij,k+1} + r_{ij,k})} \right) \times (\vec{x}_{i,k+1} + \vec{x}_{i,k} - \vec{x}_{j,k+1} - \vec{x}_{j,k}) \right\} .$$

If one then defines system work W from t_0 to t_N by

$$W = \sum_{i=1}^{n} \sum_{k=0}^{N} [(\vec{x}_{i,k+1} - \vec{x}_{i,k}) \cdot \vec{F}_{i,k}] \, ,$$

system kinetic energy K_k at time t_k by

$$K_k = \sum_{i=0}^{n} [\frac{1}{2} m_i (v_{i,k,x}^2 + v_{i,k,y}^2)] ,$$

and system potential energy V_k at time t_k by

$$V_k = \sum_{\substack{i,j=1 \\ i<j}}^{n} \left[\left(- \frac{G}{r_{ij,k}^{\alpha-1}} + \frac{H}{r_{ij,k}^{\beta-1}} \right) m_i m_j \right] ,$$

then, as in Sections 4.6 and 5.3,

$$K_N + V_N = K_0 + V_0, \quad N = 0,1,2,\ldots ,$$

which is the classical law of conservation of energy.

6.3 THE SOLID STATE BUILDING BLOCK

Let us consider next developing a viable model of a solid. In doing this, we will attempt to simulate contemporary physical thought (Feynman, Leighton, and Sands), in which molecules and atoms exhibit small vibrations within the solid. Hence, consider first a system of only two particles, P_1 and P_2, of equal mass, which interact according to (6.4). Assume that the force between the particles is zero. Then, from (6.4),

$$(6.5) \quad \frac{-G \sum_{\xi=0}^{\alpha-2} (r_{ij,k}^{\xi} \, r_{ij,k+1}^{\alpha-\xi-2})}{r_{ij,k}^{\alpha-1} \, r_{ij,k+1}^{\alpha-1} (r_{ij,k+1}+r_{ij,k})} + \frac{H \sum_{\xi=0}^{\beta-2} (r_{ij,k}^{\xi} \, r_{ij,k+1}^{\beta-\xi-2})}{r_{ij,k}^{\beta-1} \, r_{ij,k+1}^{\beta-1} (r_{ij,k+1}+r_{ij,k})} = 0.$$

But, if there is zero force between the two particles, then $r_{ij,k} = r_{ij,k+1}$, so set $r_{ij,k} = r_{ij,k+1} = r$ in (6.5) to yield

$$(6.6)\qquad \frac{-G \sum_{\xi=0}^{\alpha-2} r^{\alpha-2}}{r^{2\alpha-2}} + \frac{H \sum_{\xi=0}^{\beta-2} r^{\beta-2}}{r^{2\beta-2}} = 0.$$

Thus, for $\beta > \alpha$, which is physically reasonable (Hirschfelder, Curtis, and Bird),

$$- Gr^{-\alpha}(\alpha-1) + Hr^{-\beta}(\beta-1) = 0 ,$$

or, finally,

$$(6.7)\qquad r = \left[\frac{H(\beta - 1)}{G(\alpha - 1)}\right]^{1/(\beta-\alpha)}, \quad \beta > \alpha \geq 2 .$$

Consider next a system of only three particles, P_1, P_2 and P_3, of equal masses, and assume that no force acts between any two of the particles. Then the distance between any two of the particles is given, again, by (6.7). Such a configuration is therefore exceptionally stable and will be called a triangular building block.

When considering a solid we will decompose it into triangular building blocks. Then, by an appropriate choice of parameters, the force on any particle of a triangular block due to more distant particles will be made small, thus achieving the small vibrations desired. To illustrate, let the six particles P_1, P_2, P_3, P_4, P_5, P_6 be located at the vertices of the four triangular building blocks of the triangular region OAB, shown in Figure 6.1. Assume that $m_i \equiv 1$, $G = H = 1$, $\alpha = 7$, and $\beta = 10$, so that $r = \sqrt[3]{1.5}$. The particles' initial positions are, then,

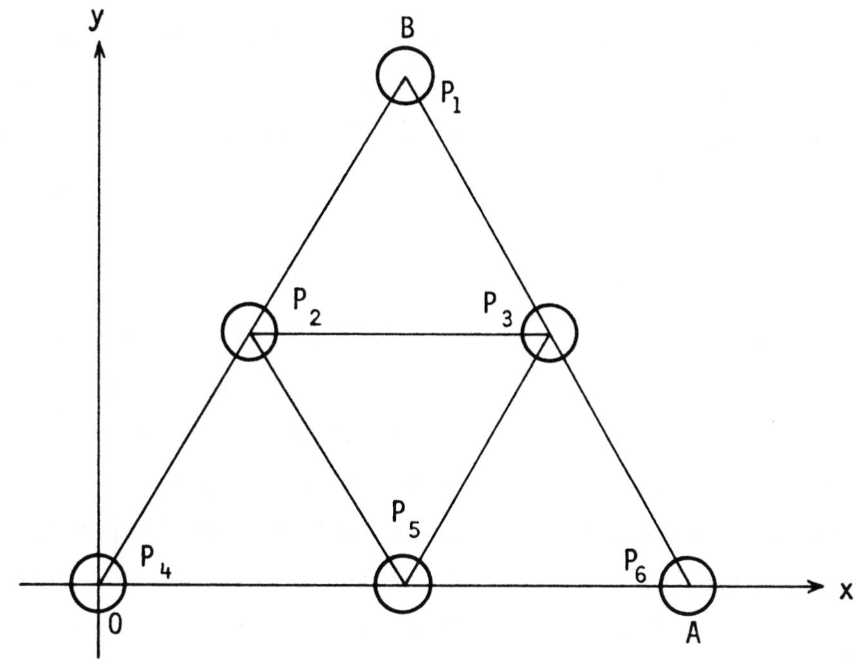

FIGURE 6.1

P_1: (1.14471, 1.98270)

P_2: (0.57236, 0.99135)

P_3: (1.71707, 0.99135)

P_4: (0, 0)

P_5: (1.14471, 0)

P_6: (2.28943, 0)

Assign to each particle a $\vec{0}$ initial velocity. Finally, let particles P_4 and P_6 be fixed and allow the remaining particles to move under force law (6.4). For $\Delta t = 0.05$ and for 2500 time steps, the motions of P_1, P_2, P_3 and P_5 were generated from (6.1)-(6.4). P_1 and P_5 exhibited small oscillations in the vertical direction only, while P_2 and P_3 exhibited small two dimensional oscillations. The maximum distance, for example, that P_1 moved from its initial position was approximately 0.02, and this occurred at approximately every one hundred time steps. The running time on the UNIVAC 1108 was 4 minutes.

6.4 FLOW OF HEAT IN A BAR

Let us now develop the basic concepts of discrete conductive heat transfer by concentrating on the prototype problem of heat flow in a bar. Physically, the problem is formulated as follows. Let the region bounded by rectangle OABC, as shown in Figure 6.2, represent a bar. Let $|OA| = a$, $|OC| = c$. A section of the boundary of the bar is heated. The problem is to describe the flow of heat through the bar.

Our discrete approach to the problem proceeds as follows. First, subdivide the given region into triangular building blocks, one such possible subdivision of which is shown in Figure 6.3 for the parameter choices $m_i \equiv 1$, $G = H = 1$, $\alpha = 7$, $\beta = 10$, $a \sim 11$, $c \sim 2$. Note that from (3.3) $r \sim 1.1447142426$.

Now, by <u>heating</u> a section of the boundary of the bar, we will mean <u>increasing the velocity</u>, and hence the potential energy, of some of the particles whose centers are on OABC. By the <u>temperature</u> $T_{i,k}$ of particle P_i at time t_k, we will mean the following. Let M be a fixed positive integer and let $K_{i,k}$ be the kinetic energy of P_i at t_k. Then $T_{i,k}$ is defined by

$$T_{i,k} = \frac{1}{M} \sum_{j=k-M+1}^{k} K_{i,j} ,$$

which is, of course, the arithmetic mean of P_i's kinetic energies at M consecutive time steps. By the <u>flow</u> of heat through the bar we will mean the transfer to other particles of the bar of the kinetic energy added at the boundary. Finally, to follow the flow of heat through the bar one need only follow the motion of each particle and, at each time step, record its temperature.

To illustrate, consider the bar shown in Figure 6.3 with the parameter choices given above, that is, $m_i \equiv 1$, $G = H = 1$, $\alpha = 7$, $\beta = 10$, $a \sim 11$, $c \sim 2$. Assume that a strong heat source is placed above P_6, and then removed, in such a fashion that $\vec{v}_{5,0} = (\frac{-\sqrt{2}}{2}, \frac{-\sqrt{2}}{2})$, $\vec{v}_{6,0} = (0,-1)$, $\vec{v}_{7,0} = (\frac{\sqrt{2}}{2}, \frac{-\sqrt{2}}{2})$, while all other initial velocities are $\vec{0}$. With regard to temperature calculation, assume that the velocities of all particles prior to t_0 were $\vec{0}$. As regards the choice of M, which is a difficult choice to make, one would usually wish to choose it relatively large, since the use of an average is, generally, more meaningful when the number of quantities being averaged is relatively large. We shall arbitrarily set $M = 20$. From the resulting calculations with $\Delta t = 0.025$, Figures 6.4-6.8 show the constant temperature contours $T = 0.1, 0.06, 0.025, 0.002$ at t_5, t_{10}, t_{15}, t_{20} and t_{25}, respectively. The resulting wave motion is clear and Figure 6.8 exhibits wave reflection. It is interesting, also, to note that the temperature at P_6 increases, until t_{20}, at which time it is a maximum, and only then does it proceed to decrease. Figures 6.9-6.13 show the constant kinetic energy contours $K = 0.1, 0.05, 0.01, 0.001$ at each of the times t_5, t_{10}, t_{15}, t_{20}, t_{25}, respectively, and indicate the magnitude of the particle velocities at these time steps.

Other heat transfer concepts can be defined now in the same spirit as above, as follows. A side of a bar is <u>insulated</u> means that the bar particles cannot transfer energy

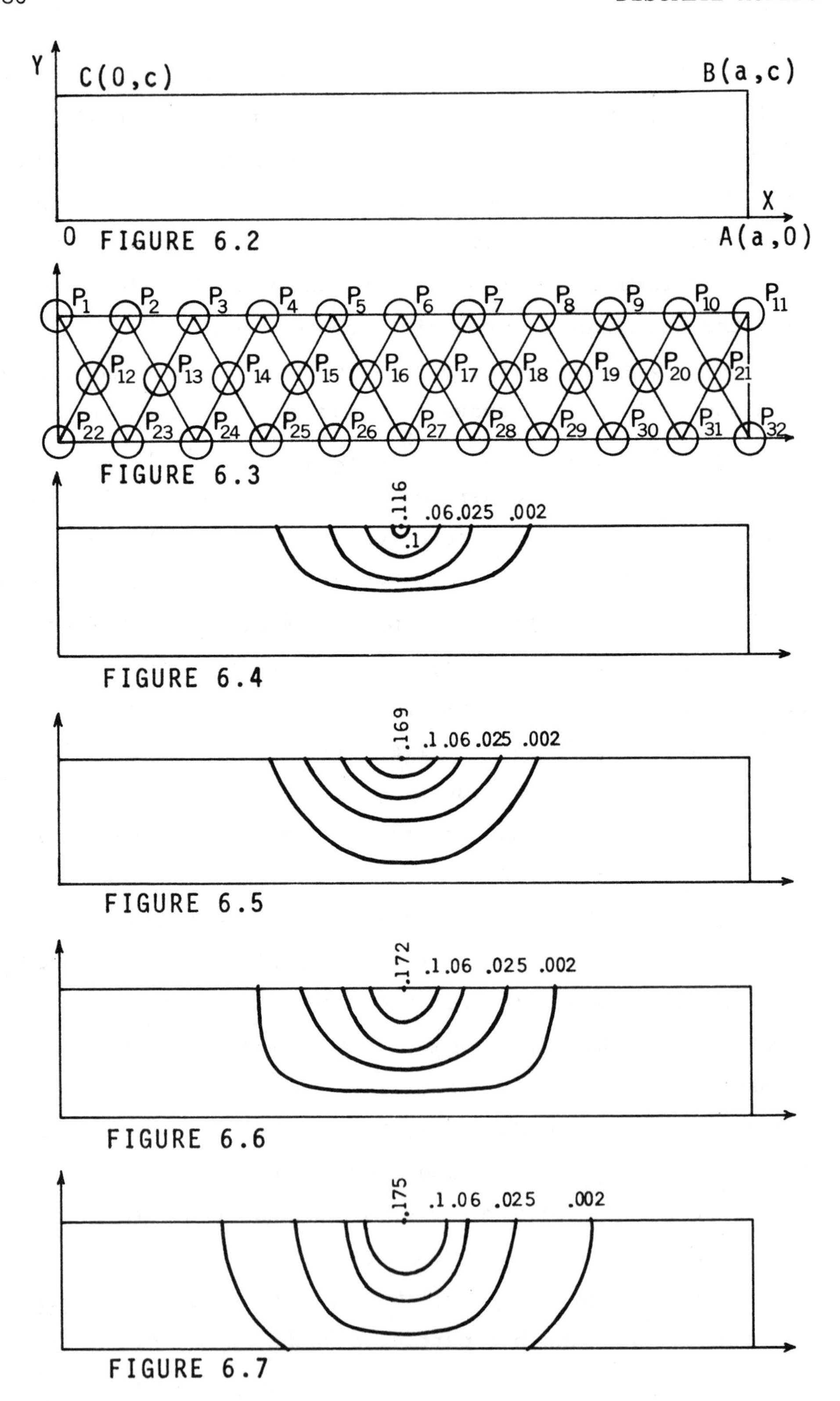

FIGURE 6.2

FIGURE 6.3

FIGURE 6.4

FIGURE 6.5

FIGURE 6.6

FIGURE 6.7

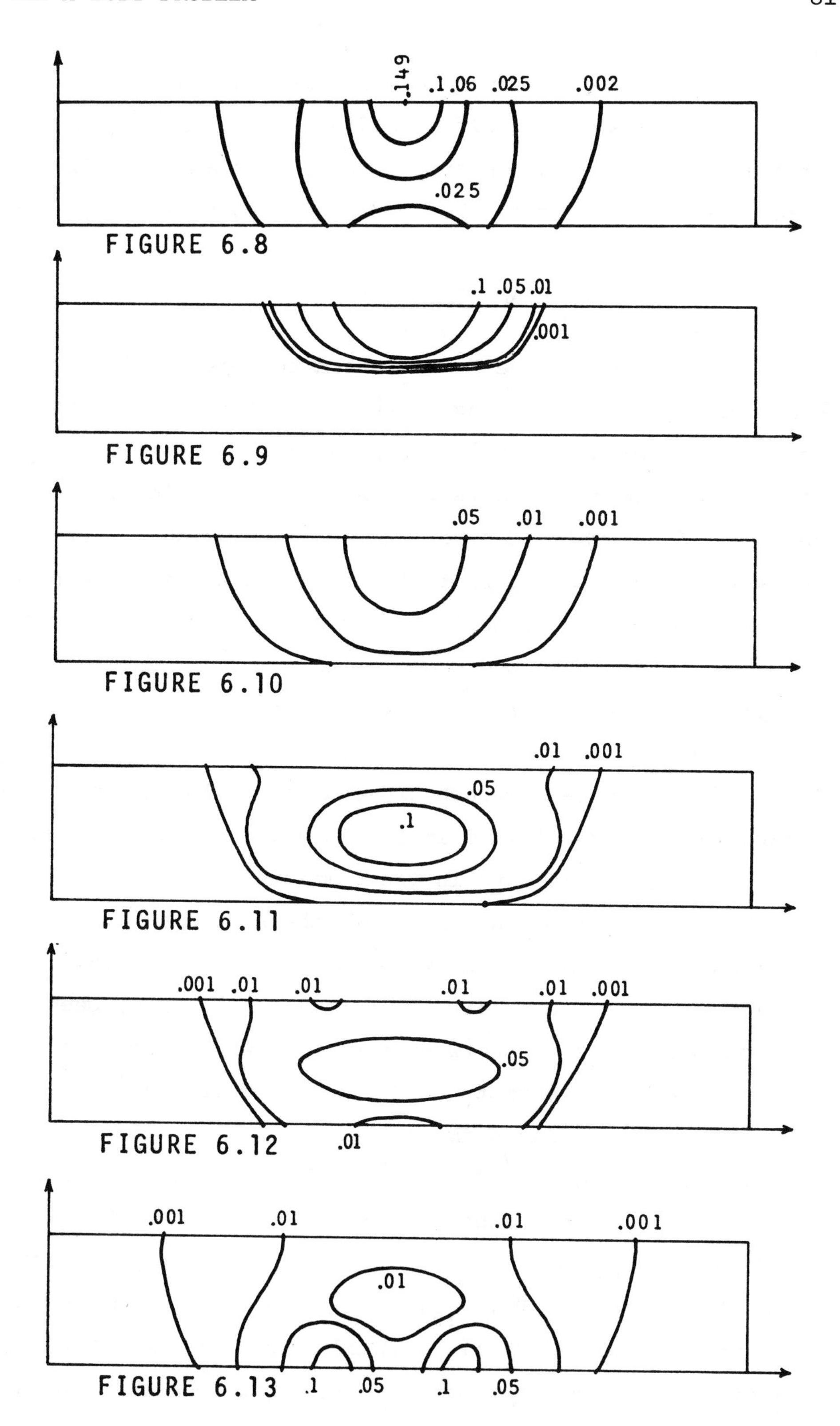

FIGURE 6.8

FIGURE 6.9

FIGURE 6.10

FIGURE 6.11

FIGURE 6.12

FIGURE 6.13

across this side of the bar to particles outside the bar, while melting is the result of adding a sufficient quantity of heat so that various particle velocities attain sufficient magnitude so as to break the bonding effect of (6.4).

6.5 OSCILLATION OF AN ELASTIC BAR

Next, let us develop the basic concepts of discrete elasticity by concentrating on the vibration of an elastic bar. The problem is formulated physically as follows. Let the region bounded by rectangle OABC, as shown in Figure 6.2, represent a bar which can be deformed, and which, after deformation, tends to return to its original shape. The problem is to describe the motion of such a bar after the external force, which has deformed the bar, is removed. Equivalently, the problem is to describe the motion of an elastic bar after release from a position of tension.

Our discrete approach proceeds as follows. The given region is first subdivided into triangular building blocks. Then, deformation results in the compression of certain particles and the stretching apart of others. Release from a position of deformation, or tension, results, by (6.4), in repulsion between each pair of particles which have been compressed and attraction between each pair which have been stretched, the net effect being the motion of the bar.

As a particular example, let $m_i \equiv 1$, $\alpha = 7$, $\beta = 10$, $G = 425$, $H = 1000$, and $\Delta t = .025$. From (6.7), $r = 1.52254$. Consider, for variety, the thirty particle bar which results by deleting P_{11} and P_{32} from the configuration of Figure 6.3. The particles P_1, P_{12}, and P_{22}, whose respective coordinates are (0, 2.63711), (.76127, 1.31855), and (0,0), are to be held fixed throughout. In order to obtain an initial position of tension like that shown in Figure 6.14a, first set P_{13}, P_{14}, P_{15}, P_{16}, P_{17}, P_{18}, P_{19}, P_{20} and P_{21} at (2.28357, 1.29198), (3.80588, 1.26541), (5.32632, 1.18573), (6.84052, 1.02658), (8.33992, .76219), (9.81058, .36813), (11.23199, -.17750), (12.57631, -.89228), and (13.80807, -1.78721), respectively. Any two consecutive points P_k, P_{k+1}, $k = 13,14,\ldots,20$, are positioned r units apart. The points $P_2 - P_{10}$ and $P_{23} - P_{31}$ are then positioned as follows: P_{k-10} and P_{k+11} are the two points which are r units from both P_k and P_{k+1} for each of $k = 12,13,\ldots,20$. Each consecutive pair of points in the $P_2 - P_{10}$ set are then separated by a distance greater than r, while each consecutive pair of points in the $P_{23} - P_{31}$ set are separated by a distance less than r. Thus, the points $P_2 - P_{10}$ are in a stretched position, while the points $P_{23} - P_{31}$ are compressed.

From the initial position of tension shown in Figure 6.14a, the oscillatory motion of the bar is determined from (6.1)-(6.4) with all initial velocities set as $\vec{0}$. The upward swing of the bar was plotted automatically at every twenty time steps and is shown in Figure 6.14a-1 from t_0 to t_{220}. It is of interest to note that as the bar moves, each row of particles exhibits wave oscillation and reflection.

6.6 REMARKS

A number of other examples were run, and these indicated that square building blocks were less stable than triangular ones, while the choices $\alpha = 2$, $\beta = 5$ and $\alpha = 7$ $\beta = 13$ were less viable than $\alpha = 7$, $\beta = 10$. Generally speaking, any choice $G > H$ resulted in increased oscillations so that, for example, for the elastic bar model of

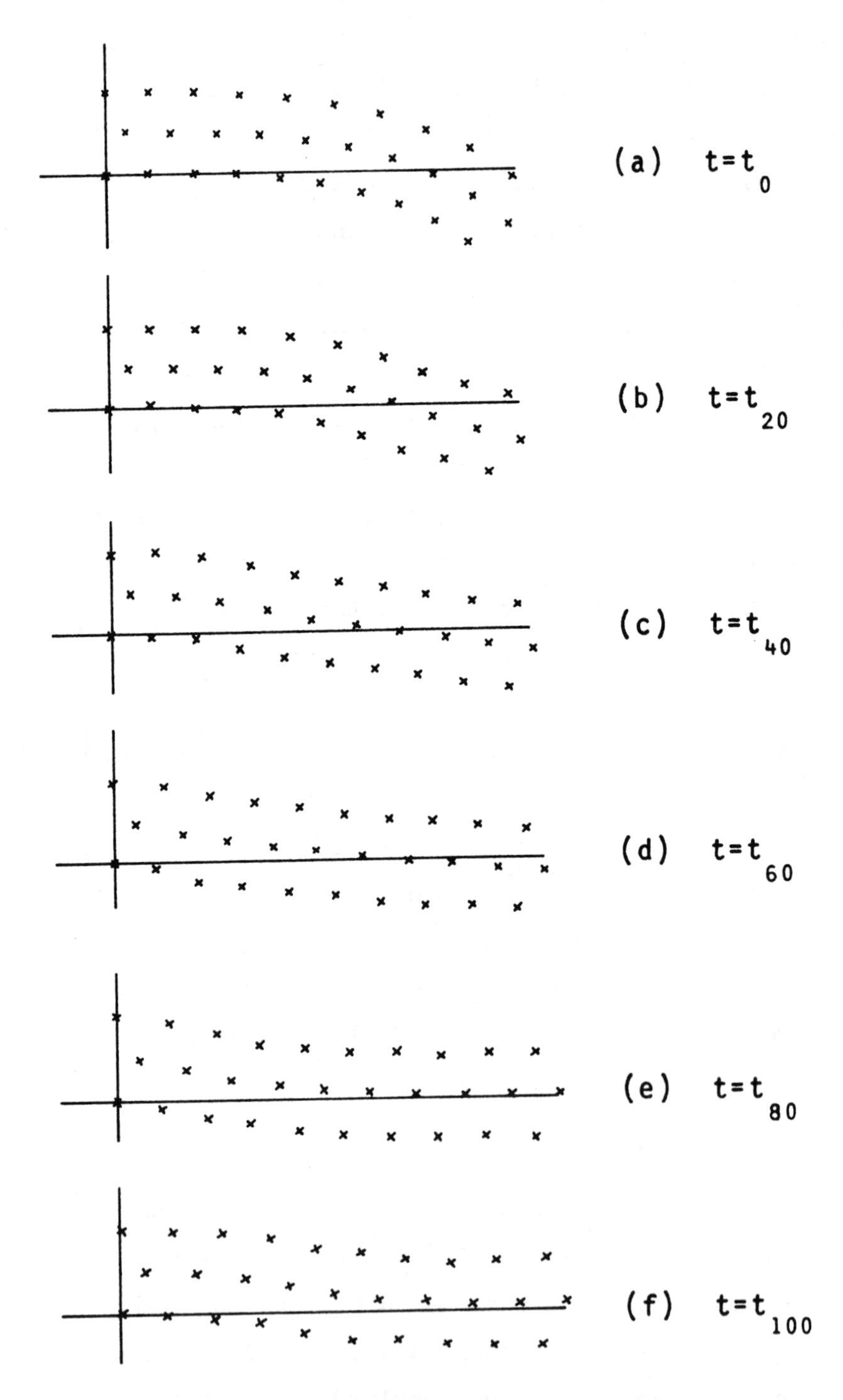

FIGURE 6.14

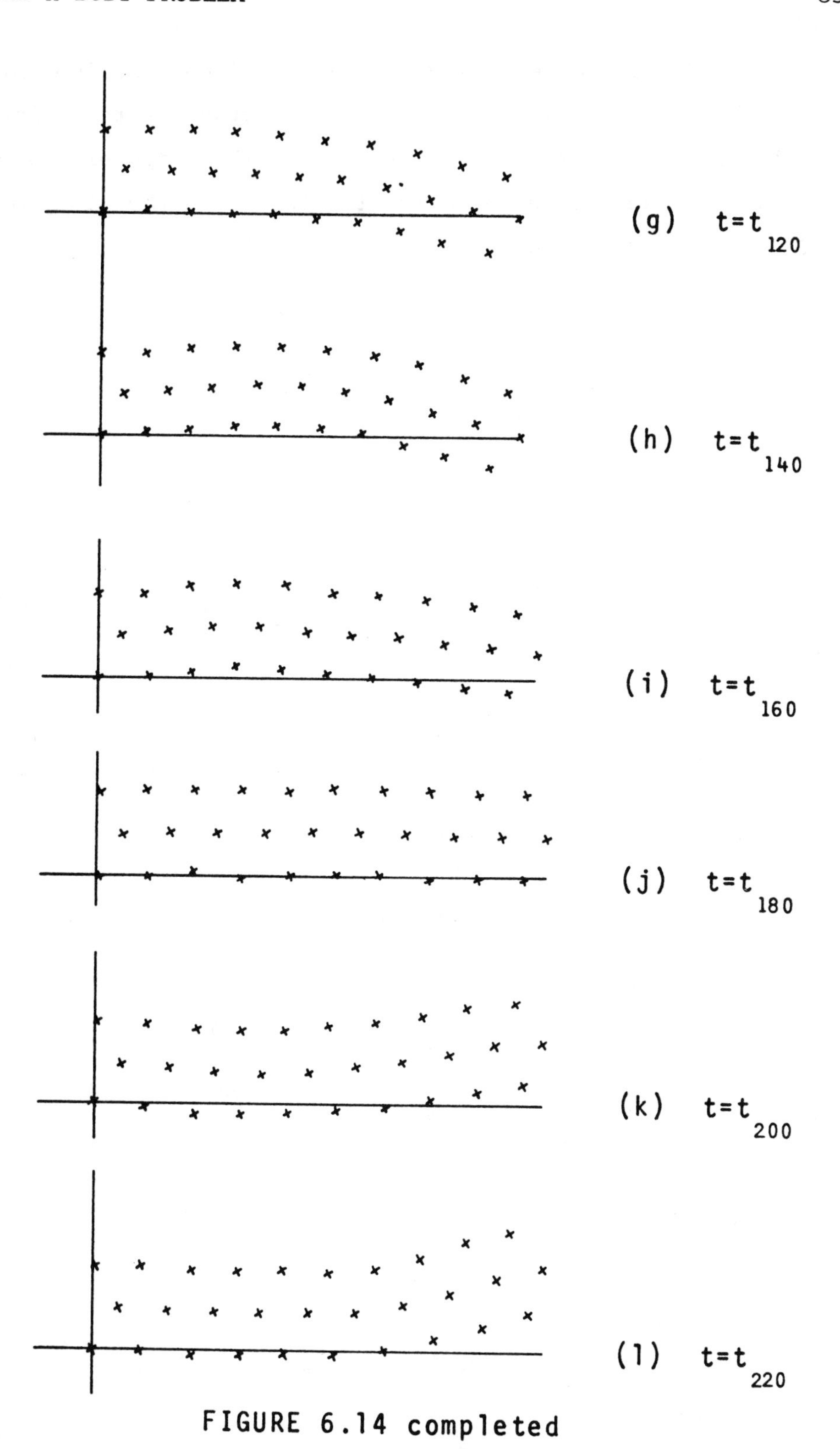

FIGURE 6.14 completed

Section 6.5 the choice $G = 3$, $H = 1$ required a refinement of time step to $\Delta t = 10^{-3}$ in order to study the resulting oscillations. The major handicap in all the computer examples run was the lack of adequate funding to enable the study of models with large numbers of particles.

CHAPTER VII - DISCRETE FLUID MODELS

7.1 INTRODUCTION

In studying fluids from a particle point of view, one could use the discrete, energy conserving formulation of Chapter VI (see, e.g., Greenspan (22)). Such an approach using relatively few particles was even recommended by von Neumann. Nevertheless, the variety of interesting fluid phenomena that can occur, like shock waves, boundary layers, and turbulence, seem to be more accessible when the related model contains a relatively large number of particles, and in terms of present generation computers, any implicit approach then becomes relatively uneconomical. For this reason, we return now to the explicit, nonconservative approach of Chapters II and III. In this chapter we will then develop a simple, "many" particle model of shock wave generation in a gas. A simple, explicit approach to turbulence will be given later, in Section 9.6.

7.2 DISCRETE SHOCK WAVES

In contrast with a liquid, a gas has relatively few particles per unit of volume. Consider, then, a gas as shown

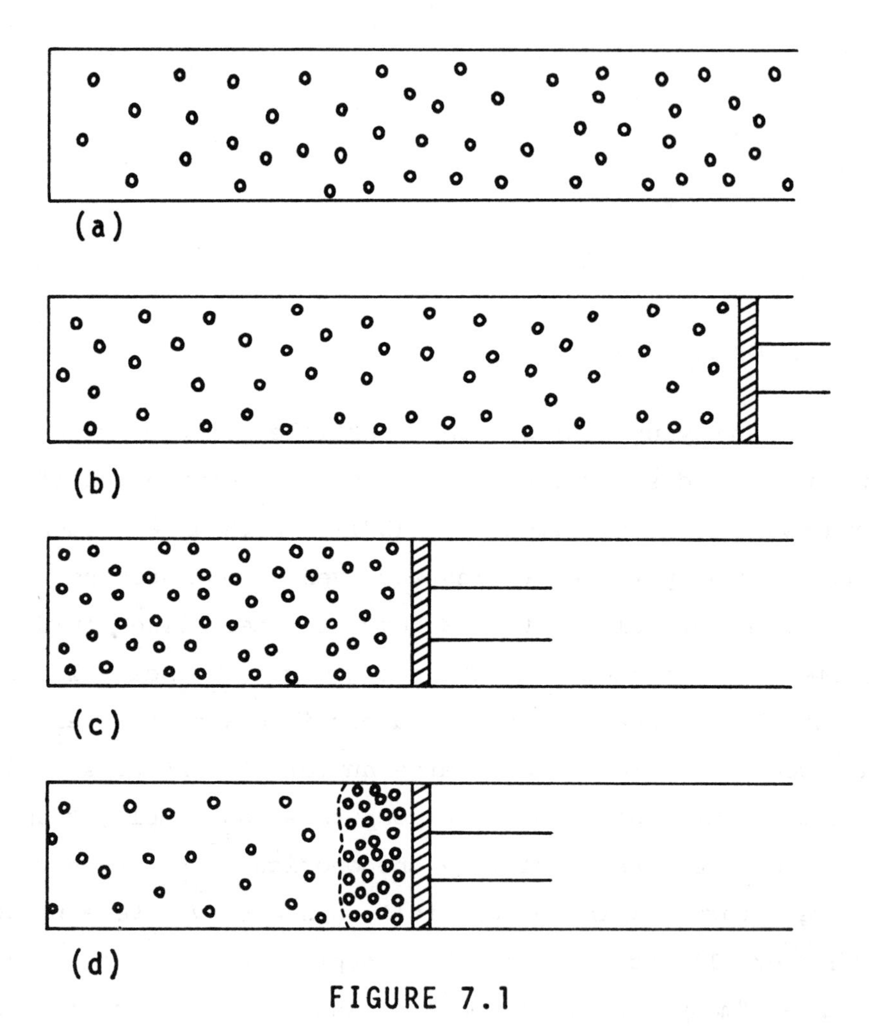

FIGURE 7.1

in a long tube in Figure 7.1a. Into this tube insert a piston, as shown in Figure 7.1b. If one first moves the piston down the tube slowly, then, as shown in Figure 7.1c, the gas particles increase in density per unit volume in a relatively uniform way. However, if, as shown in Figure 7.1d, the piston is moved at a very high rate of speed, then gas particles compact on the cylinder head, with the result that the original gas consists of two distinct portions, one with a very high density, the other with about the same density as at the start. The boundary between these two portions, which is shown as a dotted line in Figure 7.1d, is called a shock wave.

More generally, a shock wave can be thought of as follows. Assign to a given gas a positive measure of average particle density. Let a body B pass through the gas at a very high rate of speed. In certain regions about B, there may occur sets of gas particles whose densities are not average. Then, a boundary between sets of particles with average density and those with "greater than average" density is called a shock wave.

Let us illustrate this "greater than average" density concept and the development of a shock wave by considering next a particular shock tube problem. Consider the tube configuration in Figure 7.1b. For convenience, a coordinate system will be fixed relative to the piston head, as shown in Figure 7.2, so that the particles will be considered to be in motion relative to the piston. Let the tube be 100 units long, so that $AO = 100$, and 10 units high, so that $AB = 10$. Now, every $\Delta t = 0.01$ seconds, let a column of particles, each of radius $r = 0.35$ and of <u>unit</u> mass m, enter the tube at AB. Each such column is determined as follows. At each time t_k, each position $(-100, \frac{1}{2} + n)$, $n = 0,1,2,\ldots,9$, is either filled

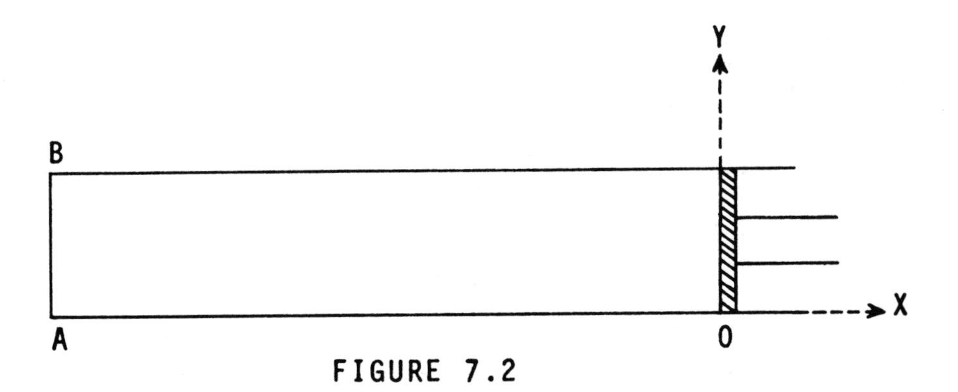

FIGURE 7.2

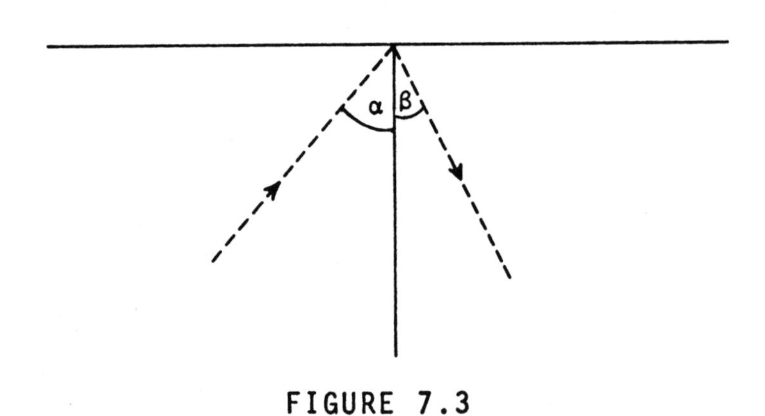

FIGURE 7.3

by a particle or left vacant by a random process, like the toss of a coin. Once it has been determined that a particle P_i is at such an initial location, then its initial velocity $\vec{v}_{i,0}$ is determined by

$$\vec{v}_{i,0} = (100 + v_{i,0,x}, v_{i,0,y}) ,$$

where $v_{i,0,x}$ and $v_{i,0,y}$ are selected at random, but are small in magnitude relative to 100, thus assuring that the gas has a relatively high speed in a relatively uniform direction. Neglecting gravitational effects, but allowing a repulsive force to simulate the effects of particle collision, we will take an n-body formulation in which the position $(x_{i,k}, y_{i,k})$ of P_i at time t_k is given explicitly by

$$(7.1) \quad x_{i,k+1} = x_{i,k} + \frac{\Delta t}{2} (v_{i,k+1,x} + v_{i,k,x})$$

$$(7.2) \quad y_{i,k+1} = y_{i,k} + \frac{\Delta t}{2} (v_{i,k+1,y} + v_{i,k,y})$$

$$(7.3) \quad v_{i,k+1,x} = v_{i,k,x} + a_{i,k,x} \Delta t$$

$$(7.4) \quad v_{i,k+1,y} = v_{i,k,y} + a_{i,k,y} \Delta t$$

$$(7.5) \quad a_{i,k,x} = \frac{1}{m} \sum_{\substack{j=1 \\ j \neq i}}^{N} \frac{(x_{i,k} - x_{j,k}) H_{ij,k}}{(r_{ij,k} + \xi)^{p+1}}$$

$$(7.6) \quad a_{i,k,y} = \frac{1}{m} \sum_{\substack{j=1 \\ j \neq i}}^{N} \frac{(y_{i,k} - y_{j,k}) H_{ij,k}}{(r_{ij,k} + \xi)^{p+1}} ,$$

where

(7.7) $r_{ij,k}$ = distance between P_i and P_j, in the tube, at t_k

(7.8) N = total number of particles in the tube at t_k

(7.9) $$H = \begin{cases} 0, & \text{if } r_{ij,k} \geq 2r \\ 1, & \text{if } r_{ij,k} < 2r , \end{cases}$$

and where p and ξ are positive parameters associated with the nature of the repulsive force between two particles which have collided. The parameter p reflects the power of the repulsion, while the parameter ξ is a measure of how close the centers of two particles can come.

If and when a particle impacts on either the top or the bottom of the tube, or on the piston head, we will assume that it rebounds, as shown in Figure 7.3, with

(7.10) $\beta = \alpha \pm \gamma$.

The quantity γ is determined at random in the range $0 \leq \gamma \leq \frac{\pi}{40}$, subject to the restriction $0 \leq \beta \leq \frac{\pi}{2}$. If this last restriction is satisfied for both choices of sign in (7.10), then the sign is to be determined at random. If the incident speed is $|v_i|$, while the reflected speed is $|v_r|$, it will be assumed that

$$|v_r| = 0.2 \; |v_i| \; ,$$

which can be interpreted as a transferrence of kinetic energy from the molecules of the gas to the molecules of the container.

For the above simple formulation with $p = 1$ and $\xi = 0.1$, Figure 7.4 shows the shock wave structure at time t_{60}, that is, after 0.6 seconds, when "greater than average" density is defined to mean that the distance between a particle and at least five other particles is less than unity.

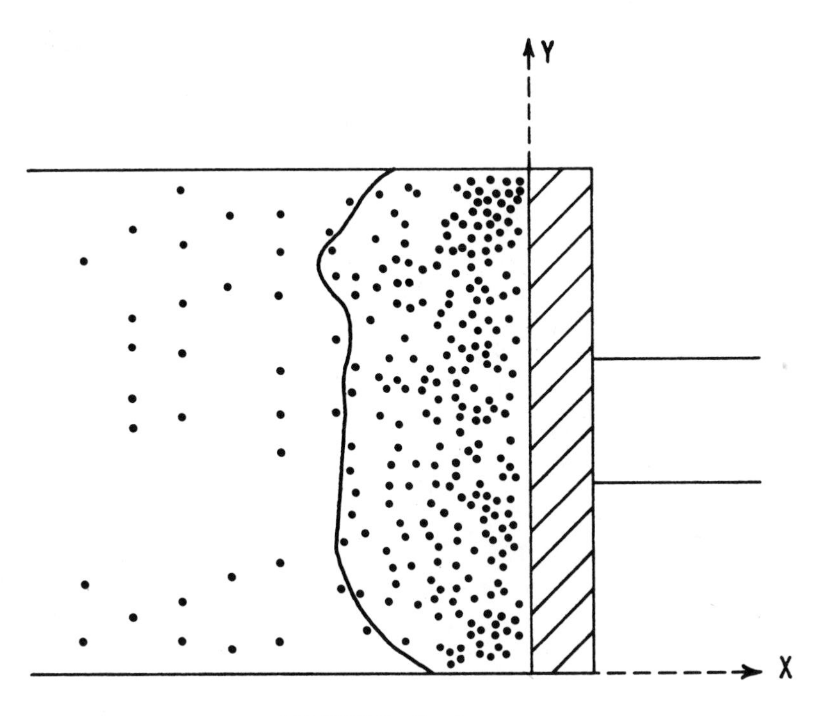

FIGURE 7.4

The points plotted are only the centers of the particles.

7.3 REMARKS

Note that an alternate discrete model approach to shock waves, which has been applied to a reasonable model of dense argon, has been developed (MacPherson), and a novel approach to the gaseous formation of a spiral galaxy, in which 300,000 particles are included, has been satisfactorily implemented (Miller, Prendergast, and Quirk). Both of these are nonconservative.

CHAPTER VIII - SYMMETRY IN DISCRETE MECHANICS

8.1 INTRODUCTION

Let us turn now to some fundamental theoretical aspects of mechanics and explore first symmetry.

The laws of Newtonian physics are invariant with respect to certain coordinate transformations. This invariance with regard to coordinates is called, in applied science, symmetry. In this chapter we will show that discrete Newtonian mechanics also enjoys the property of symmetry with regard to fundamental transformations. For clarity, we will do this in the two-dimensional context developed in Chapter IV and for convenience recall now the basic formulas of that chapter, that is,

$$(8.1) \qquad \vec{v}_k = (v_{k,x}, v_{k,y}), \qquad k = 0,1,2,\ldots,n$$

$$(8.2) \qquad \frac{v_{k+1,x} + v_{k,x}}{2} = \frac{x_{k+1} - x_k}{\Delta t}, \quad k = 0,1,2,\ldots,n-1$$

$$(8.3) \qquad \frac{v_{k+1,y} + v_{k,y}}{2} = \frac{y_{k+1} - y_k}{\Delta t}, \quad k = 0,1,2,\ldots,n-1$$

$$(8.4) \qquad \vec{a}_k = (a_{k,x}, a_{k,y}), \qquad k = 0,1,2,\ldots,n$$

(8.5) $$a_{k,x} = \frac{v_{k+1,x} - v_{k,x}}{\Delta t}, \quad k = 0,1,2,\ldots,n-1$$

(8.6) $$a_{k,y} = \frac{v_{k+1,y} - v_{k,y}}{\Delta t}, \quad k = 0,1,2,\ldots,n-1$$

(8.7) $$\vec{F}_k = m\vec{a}_k, \quad k = 0,1,2,\ldots,n .$$

(8.8) $$\vec{F}_k = (F_{k,x}, F_{k,y}), \quad k = 0,1,2,\ldots,n .$$

8.2 SYMMETRY WITH RESPECT TO TRANSLATION

Let us show first that (8.7) is invariant with respect to translation. To do this requires, essentially, proving that $\vec{a}_k$ is, in fact, a vector under the assumption that $\vec{F}_k$ is a vector (Feynman, Leighton, and Sands).

Consider then the translation

(8.9) $$x' = x - a, \quad y' = y - b ,$$

where a and b are constants. Assume $\vec{F}_k$ is a vector, so that

(8.10) $$F_{k,x} = F_{k,x'}, \quad F_{k,y} = F_{k,y'} .$$

Thus, from (8.7), (8.8), and (8.10)

(8.11) $$ma_{k,x} = F_{k,x'}$$

(8.12) $$ma_{k,y} = F_{k,y'} .$$

To complete the proof, one need only show that $a_{k,x} = a_{k,x'}$, and $a_{k,y} = a_{k,y'}$. To do this, we will show only that $a_{k,x} = a_{k,x'}$, since an analogous proof holds for the other component.

From Theorem 1.2 and equation (8.2) it follows that

$$(8.13)\quad v_{1,x} = \frac{2}{\Delta t}\,[x_1 - x_0] - v_{0,x}$$

$$(8.14)\quad v_{k,x} = \frac{2}{\Delta t}\,[x_k + (-1)^k\, x_0 + 2\sum_{j=1}^{k-1} (-1)^j x_{k-j}] + (-1)^k\, v_{0,x};\quad k \geq 2,$$

where $v_{0,x}$ is the given, x-component of $\vec{v}_0$. Define $v_{0,x'}$ by

$$(8.15)\quad v_{0,x} = v_{0,x'}\ .$$

Thus, from (8.9) and (8.13)

$$(8.16)\quad v_{1,x} = \frac{2}{\Delta t}\,[(x'_1 + a) - (x'_0 + a)] - v_{0,x'} = \frac{2}{\Delta t}\,[x'_1 - x'_0] - v_{0,x'} = v_{1,x'}\ ,$$

while from (8.9) and (8.14) for $k \geq 2$

$$(8.17)\quad v_{k,x} = \frac{2}{\Delta t}\,[(x'_k + a) + (-1)^k (x'_0 + a) + 2\sum_{j=1}^{k-1} (-1)^j (x'_{k-j} + a)] + (-1)^k v_{0,x'}\ .$$

For k odd, (8.17) yields

$$(8.18)\quad v_{k,x} = \{\frac{2}{\Delta t}\,[x'_k - x'_0 + 2\sum_{j=1}^{k-1} (-1)^j x'_{k-j}] - v_{0,x'}\} = v_{k,x'}\ ,$$

while, for k even, (8.17) yields

$$(8.19)\quad v_{k,x} = \{\frac{2}{\Delta t}\,[x'_k + x'_0 + 2a + 2\sum_{j=1}^{k-1}(-1)^j x'_{k-j}$$

$$+ 2\sum_{j=1}^{k-1}(-1)^j\, a] + v_{0,x'}\} = v_{k,x'}\ .$$

Thus, from (8.15), (8.16), (8.18) and (8.19)

$$(8.20)\quad v_{k,x} = v_{k,x'}\ ,\quad k = 0,1,2,\ldots\ .$$

Hence, from (8.5)

$$a_{k,x} = \frac{v_{k+1,x} - v_{k,x}}{\Delta t} = \frac{v_{k+1,x'} - v_{k,x'}}{\Delta t} = a_{k,x'}\ ,$$

and the proof is complete.

8.3 SYMMETRY WITH RESPECT TO ROTATION

Next, let us show that (8.7) is symmetric with respect to the rotation

$$(8.21)\quad x' = x\cos\theta + y\sin\theta$$

$$(8.22)\quad y' = y\cos\theta - x\sin\theta\ ,$$

where θ is the smallest positive angle measured in the counter-clockwise direction from the X to the X' axes. Assume again that $\vec{F}_k$ is a vector. Then, under rotation,

$$(8.23)\quad F_{k,x'} = F_{k,x}\cos\theta + F_{k,y}\sin\theta$$

$$(8.24)\quad F_{k,y'} = F_{k,y}\cos\theta - F_{k,x}\sin\theta\ .$$

Define $v_{0,x'}$ and $v_{0,y'}$ by

$$(8.25)\quad v_{0,x'} = v_{0,x}\cos\theta + v_{0,y}\sin\theta$$

$$(8.26)\quad v_{0,y'} = v_{0,y} \cos\theta - v_{0,x} \sin\theta .$$

Now,

$$(8.27)\quad v_{1,x'} = \frac{2}{\Delta t}[x_1' - x_0'] - v_{0,x'}$$

$$= \frac{2}{\Delta t}[(x_1 \cos\theta + y_1 \sin\theta)$$

$$- (x_0 \cos\theta + y_0 \sin\theta)]$$

$$- v_{0,x} \cos\theta - v_{0,y} \sin\theta$$

$$= v_{1,x} \cos\theta + v_{1,y} \sin\theta ,$$

and similarly,

$$(8.28)\quad v_{1,y'} = v_{1,y} \cos\theta - v_{1,x} \sin\theta .$$

In the same fashion as above, (8.14) implies

$$(8.29)\quad v_{k,x'} = v_{k,x} \cos\theta + v_{k,y} \sin\theta , \quad k \geq 2$$

$$(8.30)\quad v_{k,y'} = v_{k,y} \cos\theta - v_{k,x} \sin\theta , \quad k \geq 2 .$$

Finally, from (8.5), and (8.23)-(8.30), one has

$$(8.31)\quad ma_{k,x'} = m\left[\frac{v_{k+1,x'} - v_{k,x'}}{\Delta t}\right]$$

$$= [m(v_{k+1,x} \cos\theta + v_{k+1,y} \sin\theta)$$

$$- m(v_{k,x} \cos\theta + v_{k,y} \sin\theta)]/(\Delta t)$$

$$= ma_{k,x} \cos\theta + ma_{k,y} \sin\theta$$

$$= F_{k,x} \cos\theta + F_{k,y} \sin\theta$$

$$= F_{k,x'} .$$

Similarly,

$$(8.32) \quad ma_{k,y'} = F_{k,y'} \ ,$$

and the proof is complete.

8.4 SYMMETRY UNDER UNIFORM MOTION

In this section, for simplicity, we will restrict attention to one-dimensional motion. Assume then that X and X' coordinate axes are in motion relative to each other, so that

$$(8.33) \quad x' = x - c\, t_k, \ k = 0,1,2,\ldots,$$

where c is a constant. If $v_{0,x}$ is the initial velocity of a particle P on the X-axis, let its initial velocity $v_{0,x'}$ on the X'-axis be given by

$$(8.34) \quad v_{0,x'} = v_{0,x} - c \ .$$

Hence, from (8.13), (8.14), (8.33) and (8.34),

$$(8.35) \quad v_{1,x} = \frac{2}{\Delta t}\,[(x_1' + ct_1) - (x_0' + ct_0)] - v_{0,x}$$

$$= \frac{2}{\Delta t}\,[x_1' - x_0'] - v_{0,x} + 2c$$

$$= \frac{2}{\Delta t}\,[x_1' - x_0'] - (v_{0,x'}) + c$$

$$= v_{1,x'} + c \ ,$$

while, for $k \geq 2$,

$$(8.36)\quad v_{k,x} = \frac{2}{\Delta t}\{(x'_k + ct_k) + (-1)^k(x'_0 + ct_0) + 2\sum_{j=1}^{k-1}[(-1)^j(x'_{k-j}+ct_{k-j})]\} + (-1)^k v_{0,x}$$

$$= \frac{2}{\Delta t}\{x'_k + (-1)^k x'_0 + 2\sum_{j=1}^{k-1}[(-1)^j x'_{k-j}]\} + (-1)^k v_{0,x} + \frac{2c}{\Delta t}\{t_k + (-1)^k t_0 + 2\sum_{j=1}^{k-1}[(-1)^j t_{k-j}]\}.$$

But,

$$(8.37)\quad t_k + (-1)^k t_0 + 2\sum_{j=1}^{k-1}[(-1)^j t_{k-j}] = \begin{cases} 0 , & k \text{ even} \\ \Delta t , & k \text{ odd} . \end{cases}$$

Thus, from (8.34) and (8.36), for k even,

$$(8.38)\quad v_{k,x} = \frac{2}{\Delta t}\{x'_k + x'_0 + 2\sum_{j=1}^{k-1}[(-1)^j x'_{k-j}]\} + (v_{0,x'} + c)$$

$$= v_{k,x'} + c ,$$

while from (8.34) and (8.36), for k odd,

$$(8.39)\quad v_{k,x} = \frac{2}{\Delta t}\{x'_k - x'_0 + 2\sum_{j=1}^{k-1}[(-1)^j x'_{k-j}] - v_{0,x} + 2c$$

$$= v_{k,x'} + c .$$

Finally, one has from (8.35), (8.36), (8.38) and (8.39) that

$$ma_{k,x} = m\,\frac{v_{k+1,x} - v_{k,x}}{\Delta t}$$

$$= m\,\frac{(v_{k+1,x'} + c) - (v_{k,x'} + c)}{\Delta t}$$

$$= ma_{k,x'} ,$$

from which the form of Newton's dynamical equation, and hence symmetry, follow.

CHAPTER IX - OTHER FORMS OF DISCRETE MECHANICS

9.1 INTRODUCTION

It is possible to develop discrete mechanics formulations which are different from that given in Chapter I, but which are inherently more stable, and hence more economical. It is to such matters that we turn now.

9.2 AN IMPLICIT FORMULATION

For motion in a fixed X direction, let $\Delta t > 0$ and let particle P be at x_k at time $t_k = k\Delta t$, $k = 0,1,2,\ldots,n$. Let the particle's velocity $v(t_k) = v_k$ and acceleration $a(t_k) = a_k$ satisfy

$$(9.1) \qquad \frac{v_{k+1} + v_k}{2} = \frac{x_{k+1} - x_k}{\Delta t}, \quad k = 0,1,2,\ldots,n-1$$

$$(9.2) \qquad \frac{a_{k+1} + a_k}{2} = \frac{v_{k+1} - v_k}{\Delta t}, \quad k = 0,1,2,\ldots,n-1,$$

$$(9.3) \qquad ma_k = F(x_k,v_k,v_{k+1}), \qquad k = 0,1,2,\ldots,n-1.$$

where F is the force acting on P. Since (9.2) and (9.3) imply

$$(9.4)\qquad \frac{F(x_{k+1},v_{k+1},v_{k+2})+F(x_k,v_k,v_{k+1})}{2} = \frac{v_{k+1}-v_k}{\Delta t},$$

the motion of P is defined implicitly by (9.1) and (9.4).

To derive the same advantages as those described in Chapter I, one need only (Greenspan (10)) define work, with respect to (9.4), by

$$W = \sum_{k=0}^{n-1}\left[\left(\frac{F_{k+1}+F_k}{2}\right)(x_{k+1}-x_k)\right].$$

Application of (9.1) and (9.4) to the nonlinear oscillator (2.1) for each of the twelve possible cases with $v_0 = 0$; $x_0 = \frac{\pi}{4}$; $\alpha = 0.003, 0.002, 0.001, 0.000$; $\Delta t = 1, 10^{-1}, 10^{-2}$, always resulted in stability. Each time step, however, required the application of Newton's method. As a typical example, the motion for the case $\alpha = 0.003$ with $\Delta t = 10^{-2}$ is shown in Figure 9.1 up to t_{1000}.

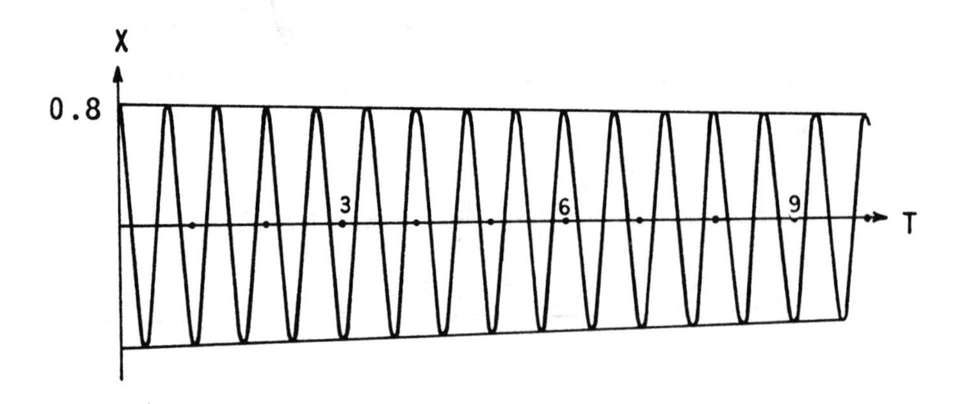

FIGURE 9.1

9.3 AN EXPLICIT FORMULATION

One can utilize another explicit approach which is somewhat different from that of Chapter II. The method is derived, heuristically, as follows.

Consider motion in a fixed X direction. For $\Delta t > 0$ let particle P be at x_k at time $t_k = k\Delta t$, $k = 0,1,2,\ldots,n$. Because of the value of central difference approximations, let us begin with the simple velocity and acceleration relationships

$$(9.5) \qquad v(t_{k+\frac{1}{2}}) = \frac{x_{k+1} - x_k}{\Delta t}, \quad k = 0,1,2,\ldots,n-1$$

$$(9.6) \qquad a(t_{k+\frac{1}{2}}) = \frac{v_{k+1} - v_k}{\Delta t}, \quad k = 0,1,2,\ldots,n-1.$$

These formulas would be completely acceptable except that $t_{k+\frac{1}{2}}$ is not a grid point in time. However, the averaging, or smoothing, approximation

$$(9.7) \qquad \frac{v_{k+1} + v_k}{2} = v(t_{k+\frac{1}{2}})$$

is also reasonable, so that (9.5) and (9.7) lead to

$$(9.8) \qquad \frac{v_{k+1} + v_k}{2} = \frac{x_{k+1} - x_k}{\Delta t}, \quad k = 0,1,2,\ldots,n-1,$$

which is, of course, nothing more than (1.5). Unfortunately, if one wishes an explicit equation of motion, the same intuition applied to acceleration will yield the method of Section 9.2. But, if one considers a_k to be the average of $a_{k+\frac{1}{2}}$ and a_{k-1}, then one must have

$$a_k = \frac{(\Delta t)\,a_{k+\frac{1}{2}} + \frac{\Delta t}{2}\,a_{k-1}}{\frac{3}{2}\,\Delta t},$$

or, equivalently,

$$(9.8) \qquad a_{k+\frac{1}{2}} = \frac{3}{2} a_k - \frac{1}{2} a_{k-1} ,$$

which, when substituted into (9.6) yields

$$(9.9) \qquad \frac{3}{2} a_k - \frac{1}{2} a_{k-1} = \frac{v_{k+1} - v_k}{\Delta t} , \quad k = 1,2,\ldots,n-1.$$

Thus, (9.9) is taken to be the basic relationship connecting velocity and acceleration. However, for $k = 1$, (9.9) contains both a_0 and a_1, so that some special procedure is, in fact, required to relate a_0 to velocity. We take this to be

$$(9.10) \qquad a_0 = \frac{v_1 - v_0}{\Delta t} ,$$

which can be called a "starting formula".

This time, in order to derive the same advantages as those described in Chapter I, we need only (Greenspan (9)) define work by

$$W = (x_1 - x_0)F(x_0,v_0,t_0) + \sum_{k=1}^{n-1} [(x_{k+1} - x_k)(\frac{3}{2} F_k - \frac{1}{2} F_{k-1})].$$

Application of formulas (9.3), (9.8), (9.9) and (9.10) to the twelve oscillator problems described in Section (9.2) yielded instability only for $\Delta t = 1$, which is an improvement over the formulas of Chapter I, but not quite as good as the formulas of Section 9.2. Those calculations which were stable yielded results almost identical to those obtained by the method of Section 9.2. The method of this section is, however, more economical than that of Section 9.2.

9.4 THE LEAP-FROG FORMULATION

As is sometimes the case, F may be independent of <u>both</u> v_k and v_{k+1} in (9.3). For such equations the formulas

$$\frac{v_{\frac{1}{2}} - v_0}{\frac{\Delta t}{2}} = a_0 \tag{9.11}$$

$$\frac{v_{k+\frac{1}{2}} - v_{k-\frac{1}{2}}}{\Delta t} = a_k \ , \quad k = 1,2,\ldots,n-1 \tag{9.12}$$

$$\frac{x_{k+1} - x_k}{\Delta t} = v_{k+\frac{1}{2}} \ , \quad k = 0,1,2,\ldots,n-1 \tag{9.13}$$

are called "leap-frog" formulas for solving

$$ma_k = F(x_k) \ , \qquad k = 0,1,2,\ldots,n-1, \tag{9.14}$$

explicitly, since they yield position at times $k\Delta t$ and velocities at times $(k + \frac{1}{2})\Delta t$. Since velocity and position are never known simultaneously, it is then somewhat meaningless even to discuss energy at any particular time. Nevertheless, the leap-frog formulas are, in practice, very stable. For example, their application to the oscillator equation (2.1) with $\alpha = 0$ and with $\Delta t = 1$, 10^{-1} and 10^{-2} was <u>always</u> stable and the results were almost identical to those obtained by the method of Section 9.2.

For variety, we shall apply these formulas next to the study of vibrating strings.

9.5 VIBRATING STRINGS, REVISITED

As described in Chapter III, calculations for the oscillations of a discrete, vibrating string became unstable easily for more than 41 particles and for $\Delta t = 0.0003$. This instability problem was improved upon by the explicit method of Section 9.3 so that one could handle up to 201 particles reasonably with $\Delta t = 0.0003$ (Greenspan (10)). But an even further economic improvement resulted if we set $\alpha = 0$ in (3.2) and used the leap-frog formulas, for then one could deal easily with 1001 particles when $\Delta t = 0.0003$ (Greenspan (8)). In this section then let us discuss the application of the leap-frog formulas to the study of string vibration under the assumption that α, in (3.2), is zero. Because not <u>all</u> the examples to be described were run for the 1001 particle string, we will restrict the discussion to a 501 particle string. We will also compare results for different types of tension formulas and see what physical consequences can be inferred.

Thus, as in (3.2), the motion of each particle P_i, $i = 1, 2,\ldots,m$, will be determined by

$$(9.15)\quad \bar{m}a_{i,k} = |T_2| \frac{(y_{i+1,k} - y_{i,k})}{[(\Delta x)^2+(y_{i+1,k}-y_{i,k})^2]^{\frac{1}{2}}} - |T_1| \frac{(y_{i,k} - y_{i-1,k})}{[(\Delta x)^2+(y_{i,k}-y_{i-1,k})^2]^{\frac{1}{2}}} - \bar{m}g\ ,$$

where $i = 1,2,\ldots,m$ and $k = 0,1,2,\ldots,n-1$, while (3.3) and (3.4) will be replaced by the leap-frog formulas

$$(9.16)\quad v_{i,\frac{1}{2}} = v_{i,0} + \frac{\Delta t}{2} a_{i,0}$$

$$(9.17)\quad v_{i,k+\frac{1}{2}} = v_{i,k-\frac{1}{2}} + (\Delta t)a_{i,k},\qquad k = 1,2,\ldots,n-1$$

$$(9.18)\quad y_{i,k+1} = y_{i,k} + (\Delta t)v_{i,k+\frac{1}{2}}\ ,\qquad k = 0,1,2,\ldots,n-1.$$

In all the examples which follow, we let $\bar{m} = 0.002$, $m = 499$, $x_0 = 0$, $x_{501} = 2$, $\Delta x = 0.004$, $\Delta t = 0.0003$, and $g = 32.2$.

Example 1. Consider first tension defined by Hooke's law, written in the particular form

$$(9.19)\quad T_2 = \frac{T_0}{\Delta x}[(\Delta x)^2 + (y_{i,k} - y_{i+1,k})^2]^{\frac{1}{2}}$$

$$(9.20)\quad T_1 = \frac{T_0}{\Delta x}[(\Delta x)^2 + (y_{i-1,k} - y_{i,k})^2]^{\frac{1}{2}},$$

where T_0 is the normalized tension between two adjacent particles whose y coordinates are equal. Substitution of (9.19) and (9.20) into (9.15) yields

$$(9.21)\quad \bar{m}a_{i,k} = \frac{T_0}{\Delta x}[y_{i+1,k} - 2y_{i,k} + y_{i-1,k}] - \bar{m}g;$$

$$i = 1,2,\ldots,m .$$

As shown in Figure 9.2, let the initial position, marked t_0, of a string for which $T_0 = 25$ be that in which the particles to the left of the line $x = 1$ are centered on $y = x$, those to the right of the line $x = 1$ are centered on $y = 1 - x$, while the center particle is positioned at (1,1). The string is released and in Figure 9.2 is shown its downward motion from t_0 to t_{960}, while in Figure 9.3 is shown its subsequent upward motion from t_{960} to t_{1920}. In Figure 9.4 is shown the motion of the middle five particles from t_0 to t_{10}. This flutter effect is omnipresent throughout the string and is the basic mechanism by which the larger motion of the string is generated.

Example 2. Example 1 was modified so that $T_0 = 50$. The result was much more rapid oscillation, with the flutter effect of the middle five particles being exhibited in Figure 9.5.

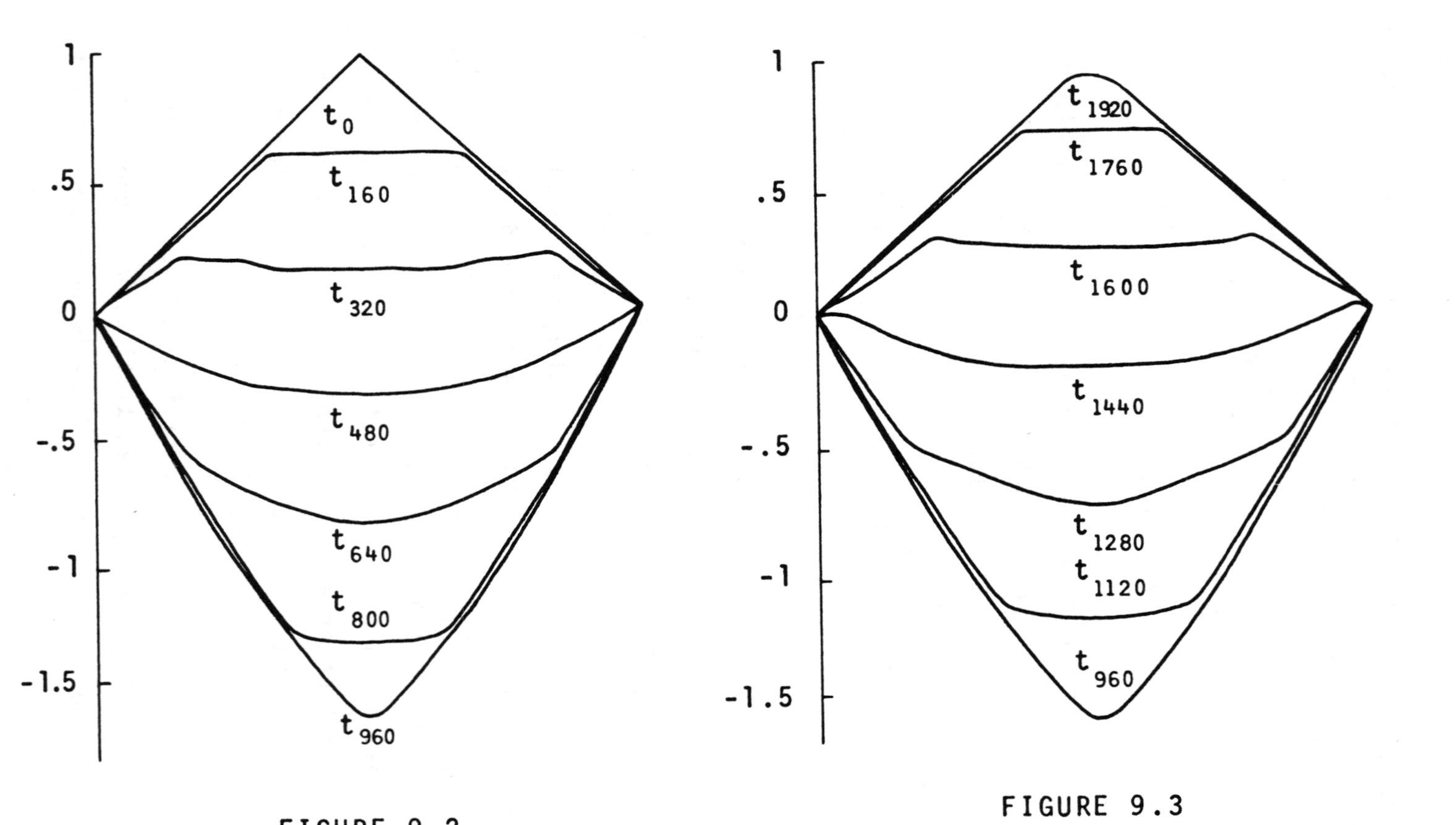

FIGURE 9.2

FIGURE 9.3

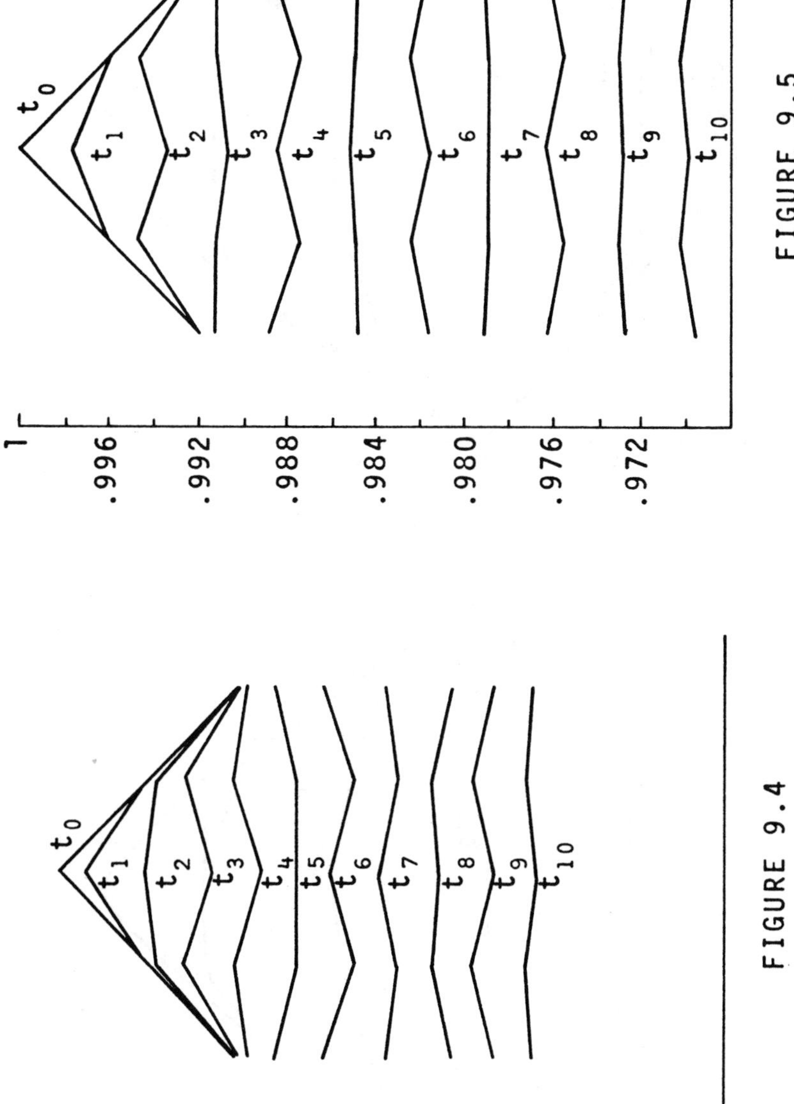

FIGURE 9.4

FIGURE 9.5

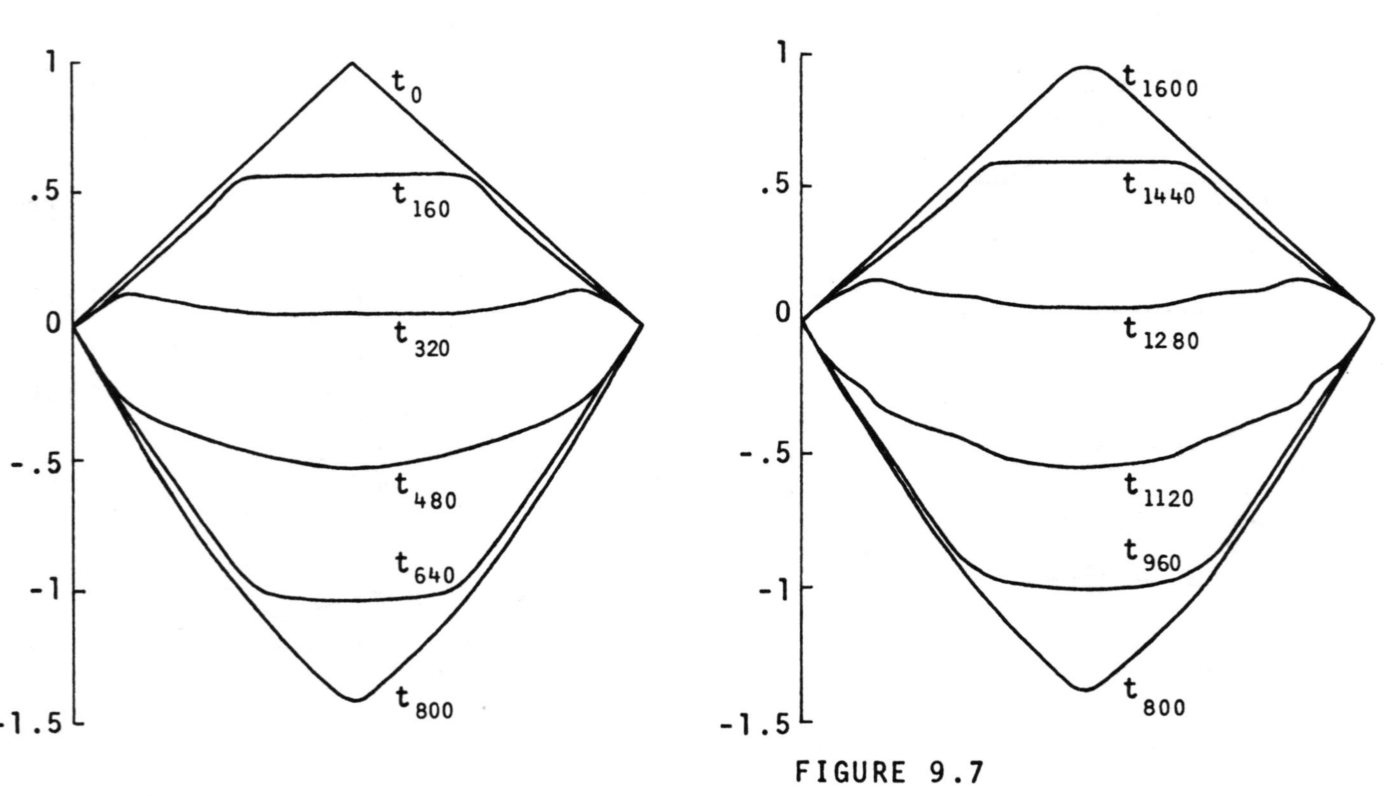

FIGURE 9.6

FIGURE 9.7

<u>Example 3</u>. Hooke's law was replaced with the tension formulas considered previously in Section 3.4, that is, by

$$(9.22)\quad T_1 = T_0 \left[1 + \left|\frac{y_{i,k}-y_{i-1,k}}{\Delta x}\right| + \varepsilon\left(\frac{y_{i,k}-y_{i-1,k}}{\Delta x}\right)^2\right]$$

$$(9.23)\quad T_2 = T_0 \left[1 + \left|\frac{y_{i+1,k}-y_{i,k}}{\Delta x}\right| + \varepsilon\left(\frac{y_{i+1,k}-y_{i,k}}{\Delta x}\right)^2\right],$$

where T_0 is, again, the normalized tension between two adjacent particles whose y coordinates are equal, and ε is a non-negative parameter. For $T_0 = 25$ and $\varepsilon = 0.01$, Figure 9.6 shows the resulting downward motion from t_0 to t_{800} of a string placed initially like that of Example 1. Figure 9.7 shows the upward motion from t_{800} to t_{1600} while Figure 9.8 shows the initial flutter motion of the five central particles. It is interesting to note, here, that the flutter seems to have a somewhat different character than that of both Examples 1 and 2. This is seen by examining the position t_4 in Figure 9.8, where the central particle is lower than both its neighbors, each of which is above the central particle but below its other neighbor. This arrangement of particles is not seen so vividly in either Figure 9.4 or Figure 9.5.

<u>Example 4</u>. Example 3 was modified by setting $\varepsilon = 0$. Though the initial motion is similar to that of Example 3, the effect of the nonlinear terms in (9.22)-(9.23) accumulates, as shown in Figure 9.9, where both results for $\varepsilon = 0.01$ and $\varepsilon = 0$ are shown on $0 \le x \le 1$ for t_{360} and t_{3600}.

<u>Example 5</u>. Example 3 was modified by choosing a new initial position with the uppermost particle at (0.2,1) and with

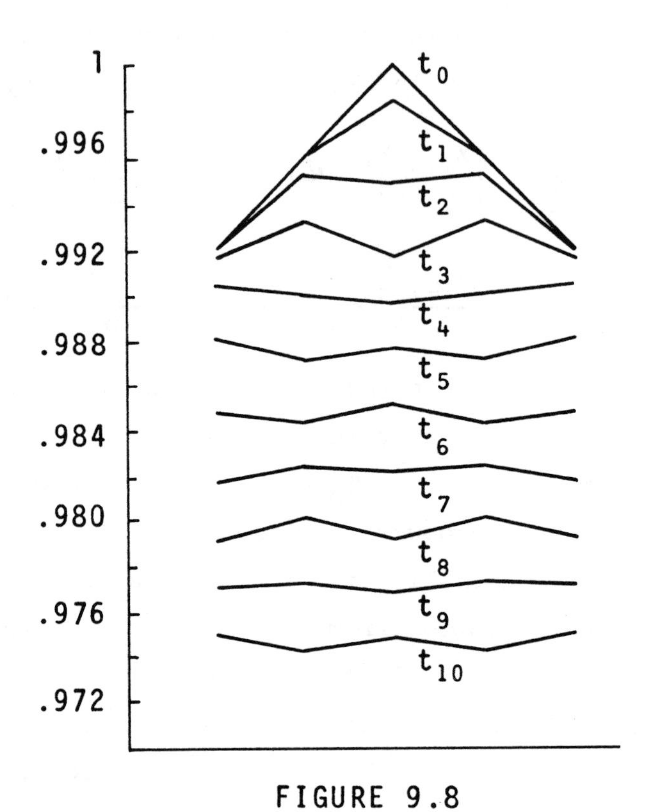

FIGURE 9.8

1.0
0
$\varepsilon=0$
$\varepsilon=10^{-2}$
t_{360}
$\varepsilon=0$
t_{3600}
$\varepsilon=10^{-2}$
-3.0

FIGURE 9.9

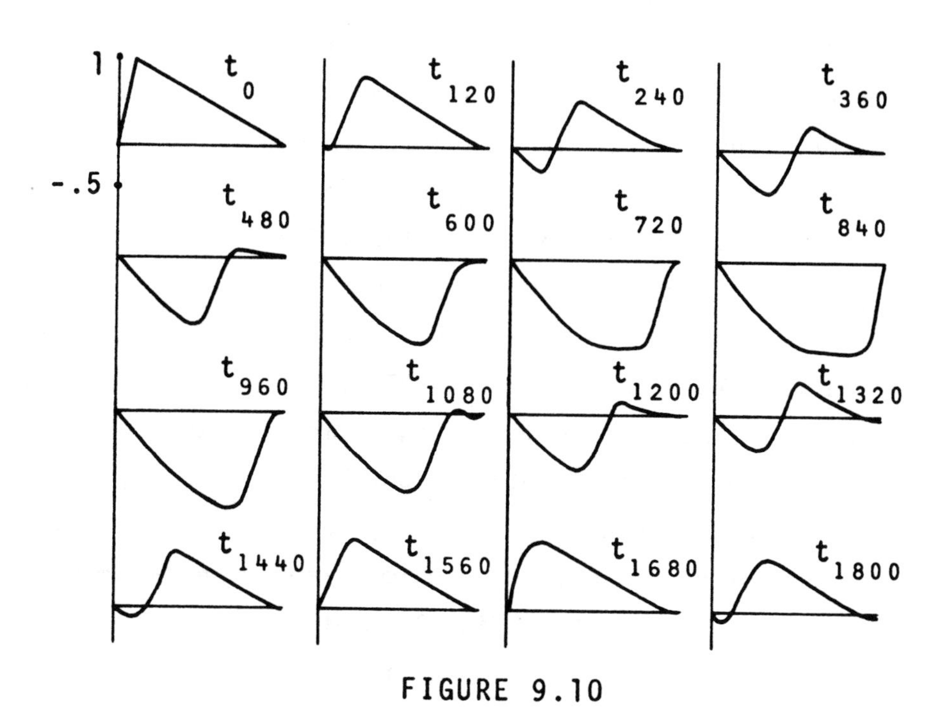

FIGURE 9.10

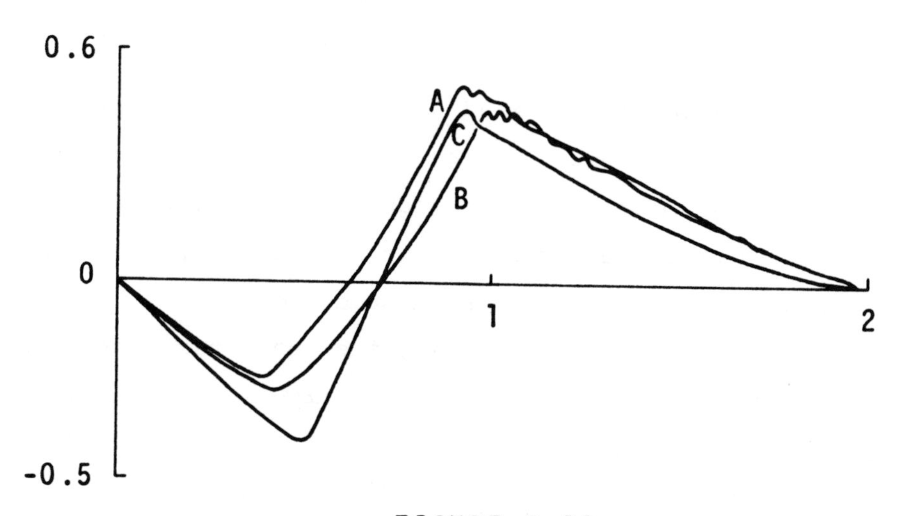

FIGURE 9.11

the remaining particles centered on the two straight lines through this point and through (0,0) and (2,0), as shown for t_0 in Figure 9.10. The resulting motion is shown through t_{1800}, when the wave is beginning its second cycle. Closer inspection of this example is quite revealing. In Figure 9.11 is shown a detailed picture, labeled A, of the string at t_{3000}. The wave labeled B is that which results at t_{3000} for $\varepsilon = 0$. The wave labeled C is that which results from Hooke's law with $T_0 = 25$ at t_{3440}. Each of A, B, and C is a second cycle configuration. Only Hooke's law gives a relatively smooth motion. The complexity of A and B could not even be achieved by increasing the tension in Hooke's law to 50, beyond which one readily encounters instability problems. Note finally that the complex wave motion shown in Figure 9.11 is the result of the initial configuration, for no such complexity appears even at t_{3600} for the symmetric strings of Figure 9.9.

9.6 THE MIXING OF FLUIDS WITH AN EXAMPLE OF DISCRETE TURBULENCE

In this section we will develop a discrete model of the mixing of two fluids. Since gravity will play a major role in the dynamical considerations, the model will be quite different in character from that in Section 7.2, and since interparticle viscosity will be assumed to be relatively unimportant, we will utilize, again, the leap-frog formulas.

Consider then a square region ABCD, as shown in Figure 9.12. For $b > 0$, let the line $y = b$ meet AD and BC in E and F, respectively. Initially, let fluid L_1 be contained in area R_1 of rectangle CDEF, while fluid L_2 is contained in area R_2 of rectangle ABFE. Under the

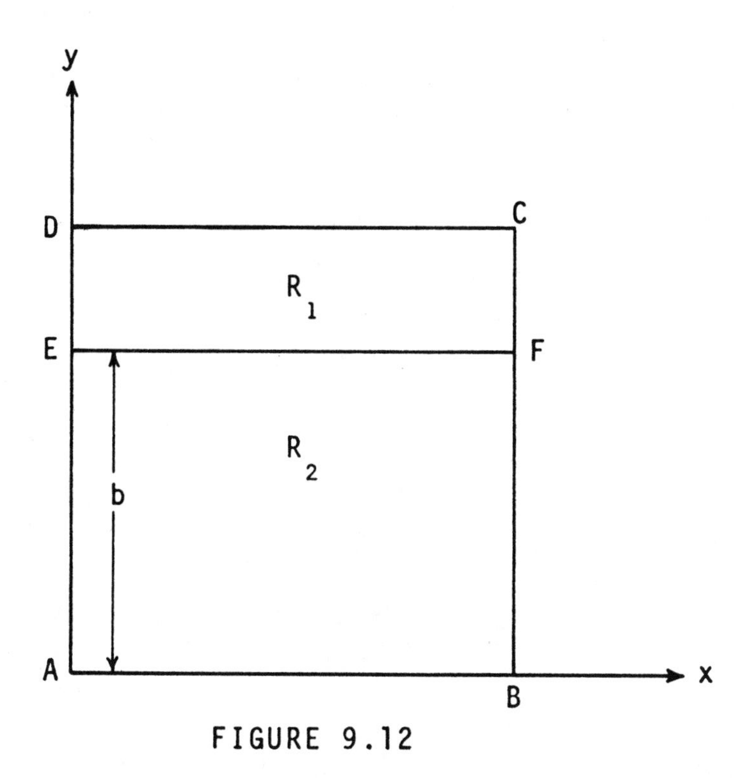

FIGURE 9.12

assumption that L_1 is more dense than L_2, the problem is to describe the resulting mixing motions of L_1 and L_2.

Let the mass of each particle of L_1 be denoted by $m(L_1)$, while that of each particle of L_2 is denoted by $m(L_2)$. Assume that

$$m(L_1) > m(L_2) \; .$$

Let L_1 consist of n particles $P_1, P_2, \ldots, P_n$, while L_2 consists of $N-n$ particles $P_{n+1}, P_{n+2}, \ldots, P_N$. If the mass of an arbitrary particle P_j is denoted by m_j, then, of course, m_j is necessarily one of $m(L_1)$ or $m(L_2)$.

Initially, let each P_j, $j = 1,2,\ldots,N$ be located at $(x_{j,0}, y_{j,0})$, have velocity $(v_{j,0,x}, v_{j,0,y})$, and have acceleration $(a_{j,0,x}, a_{j,0,y})$. For $\Delta t > 0$ and $t_k = k\Delta t$, $k = 0,1,2,\ldots$, the position $(x_{j,k}, y_{j,k})$, the velocity

$(v_{j,k+\frac{1}{2},x}, v_{j,k+\frac{1}{2},y})$, and the acceleration $(a_{j,k,x}, a_{j,k,y})$ of each P_j for each value of k are assumed to be related by

$$(9.24)\quad v_{j,k+\frac{1}{2},x} = \begin{cases} v_{j,0,x} + \frac{\Delta t}{2}(a_{j,0,x}); & k=0,\ j=1,2,\ldots,N \\ v_{j,k-\frac{1}{2},x} + (\Delta t)(a_{j,k,x}); & k=1,2,\ldots;\ j=1,2,\ldots,N, \end{cases}$$

$$(9.25)\quad v_{j,k+\frac{1}{2},y} = \begin{cases} v_{j,0,y} + \frac{\Delta t}{2}(a_{j,0,y}); & k=0,\ j=1,2,\ldots,N \\ v_{j,k-\frac{1}{2},y} + (\Delta t)(a_{j,k,y}); & k=1,2,\ldots;\ j=1,2,\ldots,N, \end{cases}$$

$$(9.26)\quad x_{j,k+1} = x_{j,k} + (\Delta t)v_{j,k+\frac{1}{2},x}; \quad j=1,2,\ldots,N$$

$$(9.27)\quad y_{j,k+1} = y_{j,k} + (\Delta t)v_{j,k+\frac{1}{2},y}; \quad j=1,2,\ldots,N\ .$$

The force $(F_{j,k,x}, F_{j,k,y})$ on P_j at t_k is assumed, as usual, to be related to the acceleration by the discrete Newtonian equations

$$(9.28)\quad m_j a_{j,k,x} = F_{j,k,x}, \qquad j=1,2,\ldots,N$$

$$(9.29)\quad m_j a_{j,k,y} = F_{j,k,y}, \qquad j=1,2,\ldots,N\ .$$

Once $F_{j,k,x}$ and $F_{j,k,y}$ are defined, then (9.24)-(9.29) determine explicitly the motion of each P_j from given initial data $x_{j,0}$, $y_{j,0}$, $v_{j,0,x}$ and $v_{j,0,y}$, $j = 1,2,\ldots,N$. Therefore, we proceed next to describe the nature of the forces to be included in the model, namely, gravity and interparticle repulsion.

For $\Delta x > 0$, if P_j is at $(x_{j,k}, y_{j,k})$ at time t_k, then the subset of particles whose centers (x,y) satisfy

$$(x_{j,k} - \Delta x) < x < (x_{j,k} + \Delta x), \quad 0 \le y < y_{j,k}$$

is called the <u>support set</u> of P_j and is denoted by $S(P_j)$. Physically, each particle in the support set of P_j is considered to be, at least in part, "beneath" P_j and thereby can contribute to preventing it from free fall. The gravitational force $g_{j,k}$ acting upon P_j at time t_k is then defined as follows.

Let K be the largest nonnegative integer such that $y_{j,k} \ge K\Delta x$ and let d_j be a positive measure of the diameter of P_j. If $K = 0$ and $y_{j,k} \le d_j$, then set $g_{j,k} = 0$, while if $K = 0$ and $y_{j,k} > d_j$, let A be the total area of $S(P_j)$ in the rectangular region defined by

$$\begin{cases} \gamma = \max\left[x_{j,k} - \frac{\Delta x}{2}, 0\right] \le x \le \min\left[x_{j,k} + \frac{\Delta x}{2}, |AB|\right] = \delta \\ 0 \le y \le y_{j,k}, \end{cases}$$

and define $g_{j,k}$ by

$$(9.30) \quad g_{j,k} = -980\left[1 - \frac{A}{(\delta-\gamma)y_{j,k}}\right].$$

If $K > 0$, consider the set of K congruent rectangles beneath P_j which are bounded by

$$\begin{cases} \gamma = \max\left[x_{j,k} - \frac{\Delta x}{2}, 0\right] \le x \le \min\left[x_{j,k} + \frac{\Delta x}{2}, |AB|\right] = \delta \\ y = p\Delta x; \quad p = 0,1,2,\ldots,K. \end{cases}$$

If the intersection of any one of these rectangles with $S(P_j)$

is empty, set $g_{j,k} = -980$. But, if each of these rectangles has a nonzero intersection with $S(P_j)$ and if A is the total area of these nonzero intersections, then set

$$(9.31)\quad g_{j,k} = -980 \left[1 - \frac{A}{(\delta-\gamma)K\Delta x}\right].$$

From (9.31), note that if $A = (\delta-\gamma)K\Delta x$, then P_j has no gravity acting upon it, that is, it is supported fully by particles below it. However, if $A < (\delta-\gamma)K\Delta x$, then $g_{j,k}$ is proportional to how much support is below P_j, while if $A > (\delta-\gamma)K\Delta x$ then many particles have been compressed beneath P_j and the resulting force will be antigravitational. Similar conclusions hold with respect to (9.30)

To simulate repulsion between the particles, whether it be due to collision or electrical forces, we will proceed now as follows. Let P_i have mass m_i and be located at $(x_{i,k}, y_{i,k})$ at time t_k. Let P_j have mass m_j and be located at $(x_{j,k}, y_{j,k})$ at time t_k. Let $r_{ij,k}$ be the distance between $(x_{i,k}, y_{i,k})$ and $(x_{j,k}, y_{j,k})$. Then, the force of repulsion on P_j exerted by P_i is defined by

$$(9.32)\quad F_{j,k,x} = \frac{\alpha m_i m_j (x_j - x_i)}{(r_{ij,k}+\xi)^p \left(\dfrac{r_{ij,k}+\xi}{d_j}\right)^\beta},$$

$$F_{j,k,y} = \frac{\alpha m_i m_j (y_j - y_i)}{(r_{ij,k}+\xi)^p \left(\dfrac{r_{ij,k}+\xi}{d_j}\right)^\beta},$$

where α is a nonnegative constant, ξ is a positive measure of how close the centers of two particles are allowed to be, p is a positive exponent of repulsion, and β is a nonnegative exponent of repulsion which is zero except when

$r_{ij,k} < d_j$, at which time it is positive. The effect of β is to greatly increase the force of repulsion when two particles are exceptionally close, as when, for example, they have collided.

The equations of motion of each P_j are then defined by (9.24)-(9.26) with

$$(9.33)\quad a_{j,k,x} = \sum_{\substack{i=1\\ i\neq j}}^{N} \frac{\alpha m_i (x_j - x_i)}{(r_{ij,k}+\xi)^p \left(\frac{r_{ij,k}+\xi}{d_j}\right)^{\beta}}, \quad j = 1,2,\ldots,N$$

$$(9.34)\quad a_{j,k,y} = \left\{ \sum_{\substack{i=1\\ i\neq j}}^{N} \frac{\alpha m_i (y_j - y_i)}{(r_{ij,k}+\xi\;^{p} \left(\frac{r_{ij,k}+\xi}{d_j}\right)^{\beta}} \right\} + g_{j,k}, \quad j = 1,2,\ldots,N.$$

Collision at the wall will be treated simply by assuming that the angle of incidence is the same as the angle of reflection, and that the reflected speed v_r is related to the incidence speed v_i by

$$(9.35)\quad |v_r| = \omega |v_i|, \quad 0 < \omega \leq 1 .$$

The initial velocities of $P_1, P_2, \ldots, P_N$ will be determined as random quantities in the ranges

$$- V \leq v_{j,0,x} \leq V$$

$$- V \leq v_{j,0,y} \leq V ,$$

where V is a fixed positive constant.

From the large number of examples run, let us describe now only two which are both typical and physically reasonable. In each case, the choices $|AB| = 100$ and $b = 75$ were used for the square shown in Figure 9.12.

<u>Example 1</u>. Consider a sixteen particle configuration with $n = 4$, $N = 16$, $m(L_1) = 25$, $m(L_2) = 10$, $\Delta x = 25$, $\Delta t = 10^{-3}$, $d_j \equiv d = 20$, $\alpha = 1$, $\xi = 0$, $p = 2$, $\beta = 5$, $\omega = 1$, and $V = 100$. The initial positions and initial velocities were

(12.5, 87.5) ,	$v_x = -2.90$,	$v_y = 48.91$
(37.5, 87.5) ,	$v_x = 98.43$,	$v_y = -87.83$
(62.5, 87.5) ,	$v_x = 22.21$,	$v_y = 41.44$
(87.5, 87.5) ,	$v_x = -27.61$,	$v_y = 47.97$
(12.5, 62.5) ,	$v_x = 46.99$,	$v_y = -1.53$
(37.5, 62.5) ,	$v_x = -99.65$,	$v_y = -16.10$
(62.5, 62.5) ,	$v_x = 26.14$,	$v_y = 80.82$
(87.5, 62.5) ,	$v_x = -42.70$,	$v_y = 56.75$
(12.5, 37.5) ,	$v_x = 27.02$,	$v_y = -75.35$
(37.5, 37.5) ,	$v_x = 98.66$,	$v_y = -69.15$
(62.5, 37.5) ,	$v_x = 93.48$,	$v_y = -85.65$
(87.5, 37.5) ,	$v_x = -46.35$,	$v_y = 88.20$
(12.5, 12.5) ,	$v_x = -9.09$,	$v_y = 90.73$
(37.5, 12.5) ,	$v_x = -6.56$,	$v_y = 95.21$
(62.5, 12.5) ,	$v_x = -13.07$,	$v_y = -49.69$
(87.5, 12.5) ,	$v_x = 50.50$,	$v_y = -13.84$.

Figures 9.13-9.21 show the relative positions of L_1 and L_2 at the consecutive times $t_{160+240k}$, $k = 0,1,2,3,\ldots,8$. The particles of L_1 are labeled A while those of L_2 are labeled B. Figure 9.21 shows the complete interchange of the relative positions of L_1 and L_2, so that the "heavier" fluid has settled to the bottom. Thereafter, until t_{3040}, all the particles continue to be in motion, but at least three from L_1 always remain at the bottom.

<u>Example 2</u>. Consider a 256 particle configuration with

$n = 64$, $N = 256$, $m(L_1) = 1$, $m(L_2) = 0.25$, $\Delta x = 6.25$, $\Delta t = 10^{-2}$, $d_j \equiv d = 5$, $\xi = 0.1$, $p = \beta = 2$, $\omega = 0.9$, $V = 500$. The computation of acceleration components (9.33) and (9.34) was simplified by assuming that each particle was acted upon only by "nearby" particles. This was implimented by defining α as follows:

$$\alpha = \begin{cases} 0 , & \text{if } r_{ij,k} \geq 2d = 10 \\ 1 , & \text{if } r_{ij,k} < 2d = 10 . \end{cases}$$

The initial positions of the particles were fixed at the points $(3.125 + 6.25\mu_1, 3.125 + 6.25\mu_2)$, $\mu_1 = 0,1,2,\ldots,15$, $\mu_2 = 0,1,2,\ldots,15$.

Figure 9.22 shows the initial interaction between L_1 and L_2 at t_2. The particles of L_1 are labeled A, while those of L_2 are labeled B. An arbitrary boundary has been drawn between L_1 and L_2 to indicate the type of motion in progress. Figure 9.23 shows the state of diffusion at time t_{69} by the setting of circles around the particles of L_1. This diffuse character persisted during the entire calculation. At approximately $t = t_{250}$, the small damping effect incorporated in (9.35) became evident in that the particle velocities had decreased noticeably. Figure 9.24, then, shows not only the state of diffusion at time t_{300}, but also shows the onset of a "thinning" of particles in the upper portion of the region and a "condensation" of particles in the lower portion, due probably to a resultant loss of energy in the system. Figure 9.25 shows the position, and resultant motion, at times t_{10k}, $k = 0,1,2,\ldots,30$, for the particle whose initial position was (53.1, 46.9) and whose initial velocity components, generated, of course, at random, were

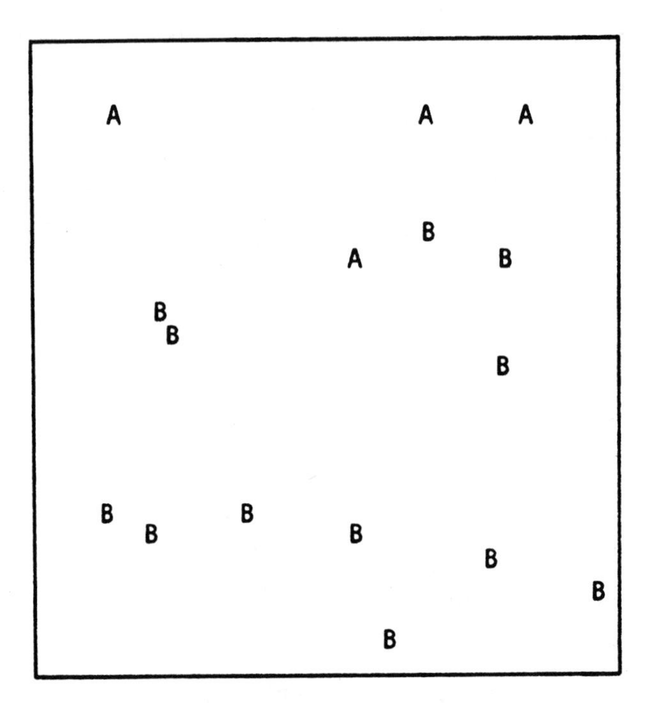

FIGURE 9.13. $T=t_{160}$

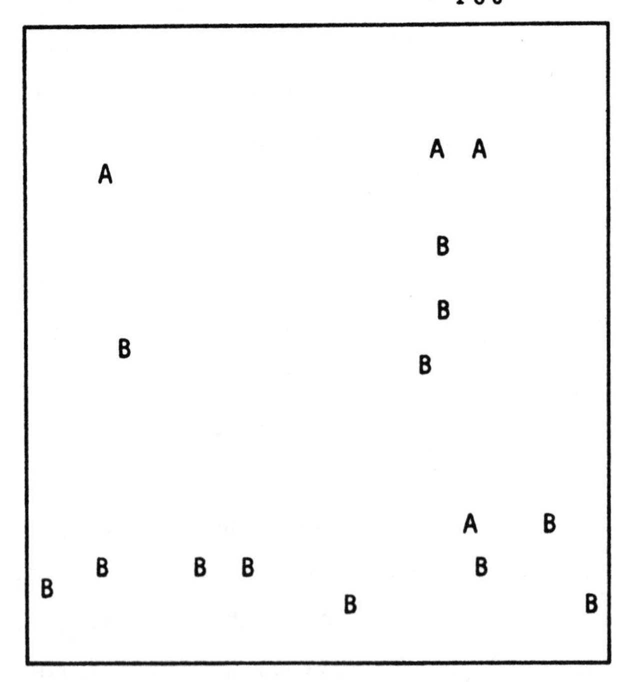

FIGURE 9.14. $T=t_{400}$

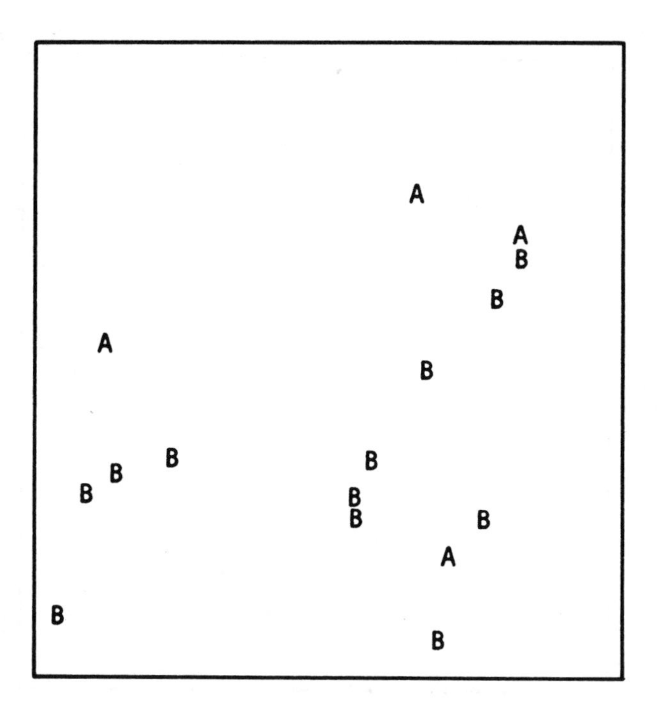

FIGURE 9.15. $T=t_{640}$

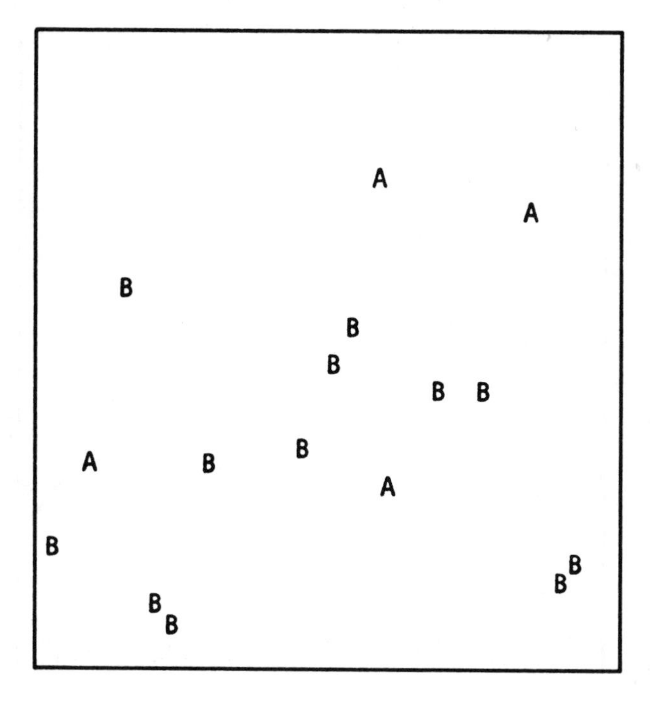

FIGURE 9.16. $T=t_{880}$

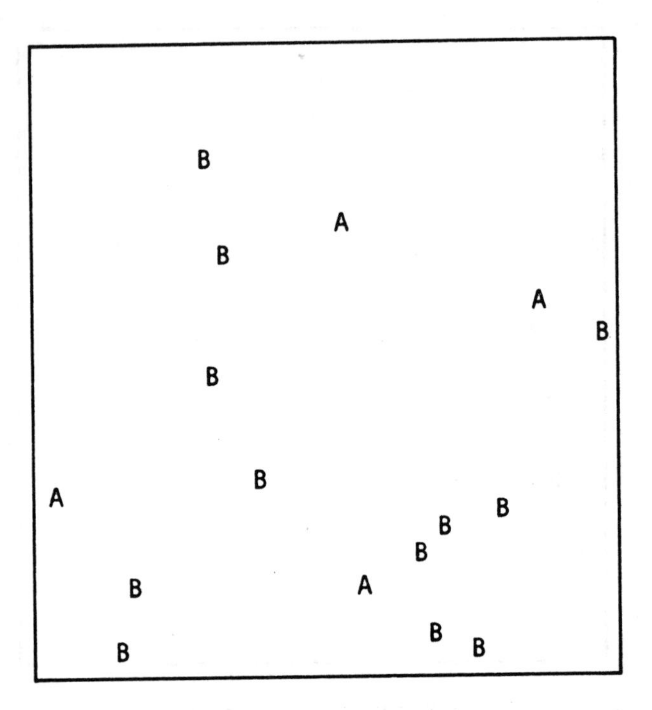

FIGURE 9.17. $T=t_{1120}$

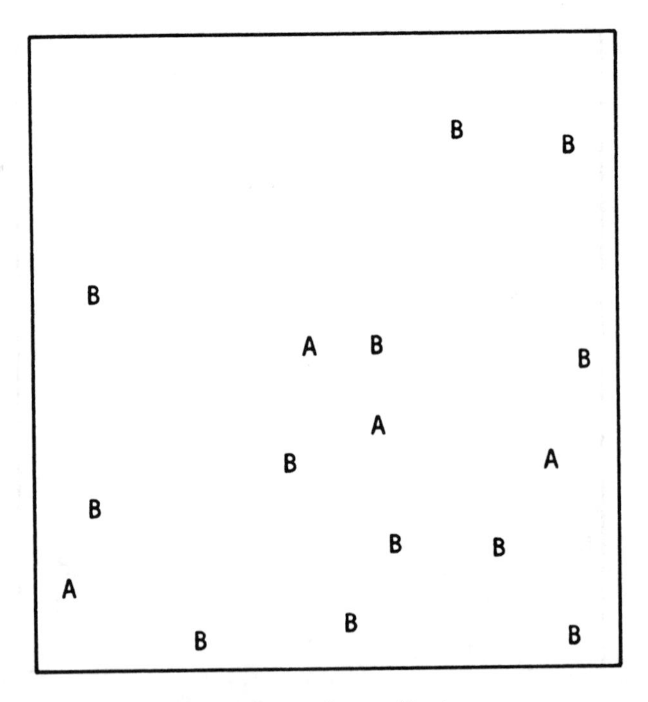

FIGURE 9.18. $T=t_{1360}$

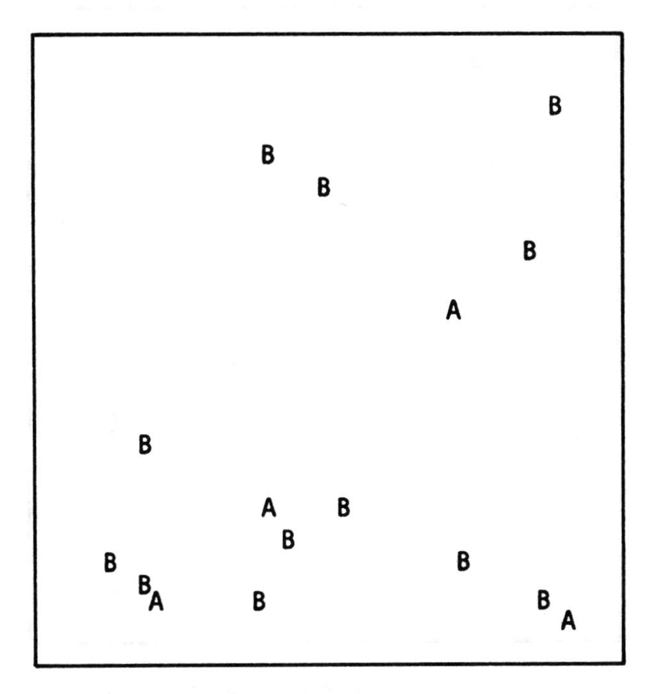

FIGURE 9.19. $T=t_{1600}$

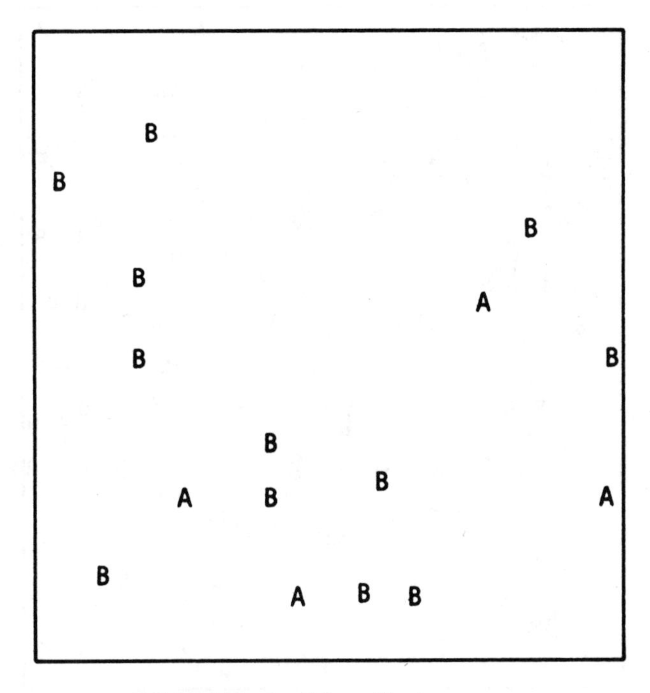

FIGURE 9.20. $T=t_{1840}$

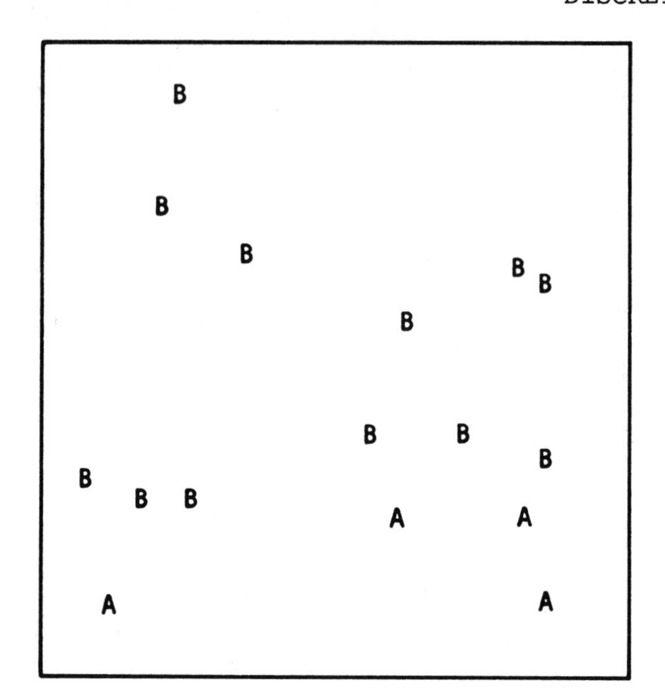

FIGURE 9.21. $T=t_{2080}$

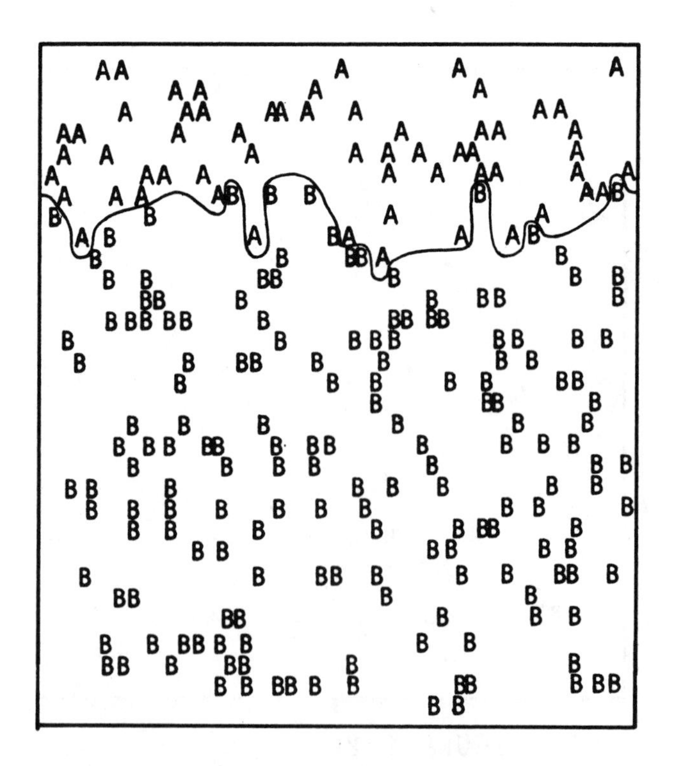

FIGURE 9.22. $T=t_2$

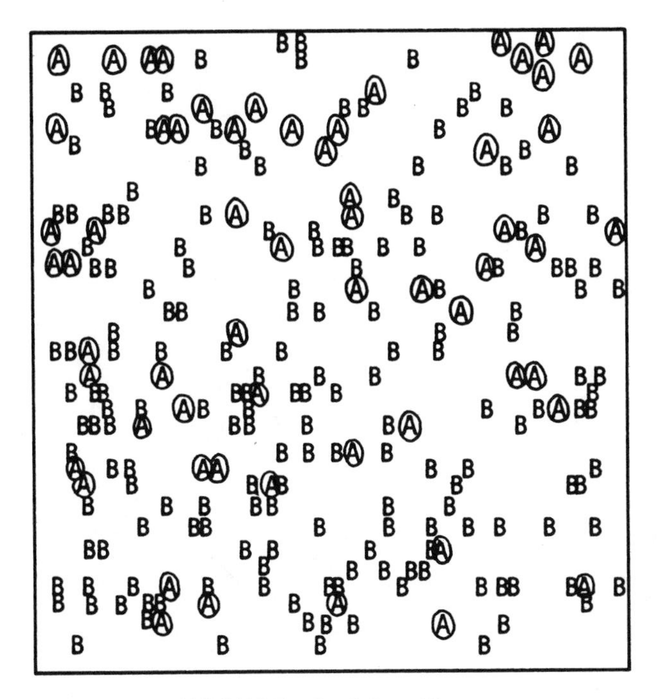

FIGURE 9.23. $T=t_{69}$

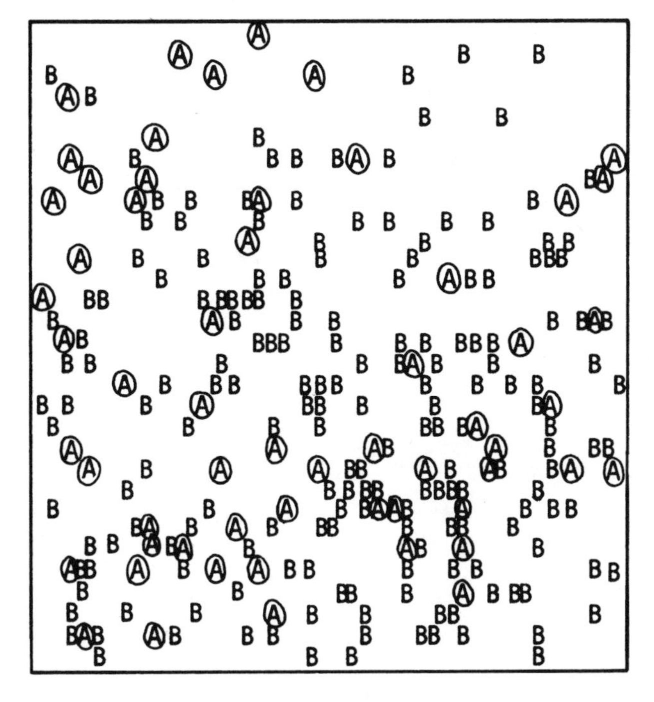

FIGURE 9.24. $T=t_{300}$

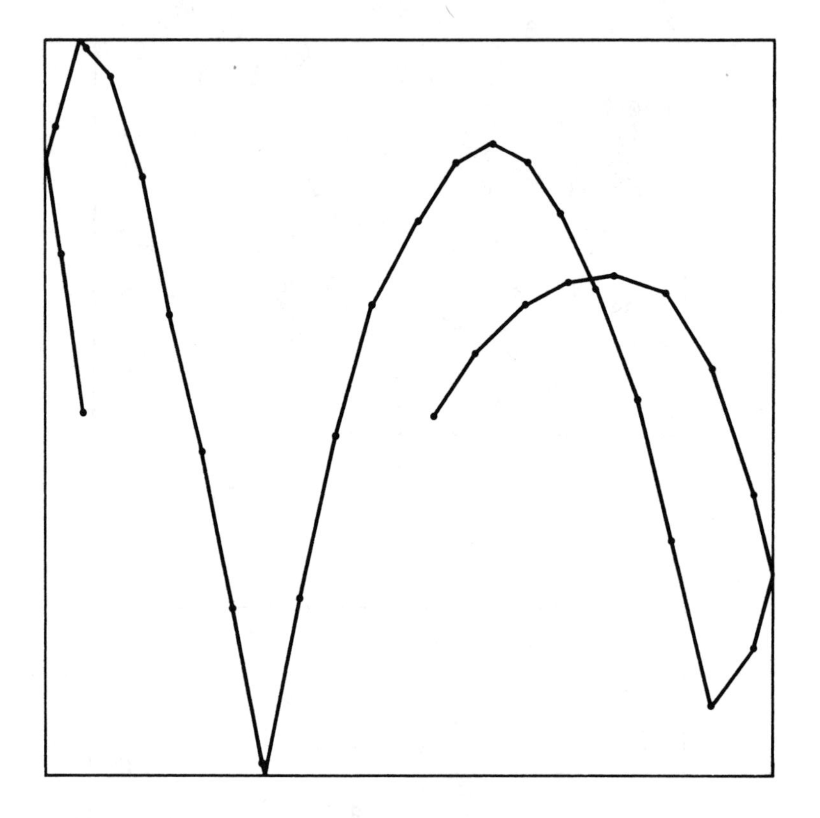

FIGURE 9.25

$v_x = 63.0$, $v_y = 133.9$. The figure not only shows the strong effect of gravity, but, as can be seen from the lower right hand corner, also shows the strong effect of repulsion. Figure 9.26 shows the vector field defined by the directions of the particles at time t_{92}. If one interprets the indicated oval areas, with respect to which all nearby vectors have the same relative orientation, as vortices, then these patterns of vortices are in a constant state of formation and decay, due to the relatively independent motion of each particle. This rapid appearance and disappearance of small vortices is, of course, a fundamental characteristic of turbulent motion.

It should be noted that automatic graphing limitations did not allow for the superposition of letters so that, in Figures 9.22-9.24, when two particles were exceptionally close, only one letter, A or B, was printed. In cases where a choice between A and B had to be made, A was always printed.

Finally, note that, for Example 2, the total running time up to t_{300}, that is, for 300 time steps, was only 72 minutes.

With regard to the method and examples of this section, other examples indicated that the character of the fluid motion could be changed in an expected fashion by appropriate variations of the parameters of the equations of motion. Thus, for example, an increase in d invariably led to a decrease in the rate of diffusion, while an increase in β could lead to instability if Δt was not decreased simultaneously. If one wished to choose Δt sufficiently small, then, indeed, one could actually set $d = 1$ and choose p and β so as to agree with the repulsive part of any of the commonly accepted molecular potential functions (Hirschfelder, Curtis, and Bird; MacPherson).

9.7 REMARKS

Though the implicit, explicit, and leap-frog formulations of Sections 9.2-9.4 were derived intuitively by appealing to the desirability of both smoothing and averaging, these formulas can be constructed directly, in the usual way, by means of finite Taylor expansions (see, e.g., Greenspan (6)). The validity of such an approach follows from the discussion of Section 1.1.

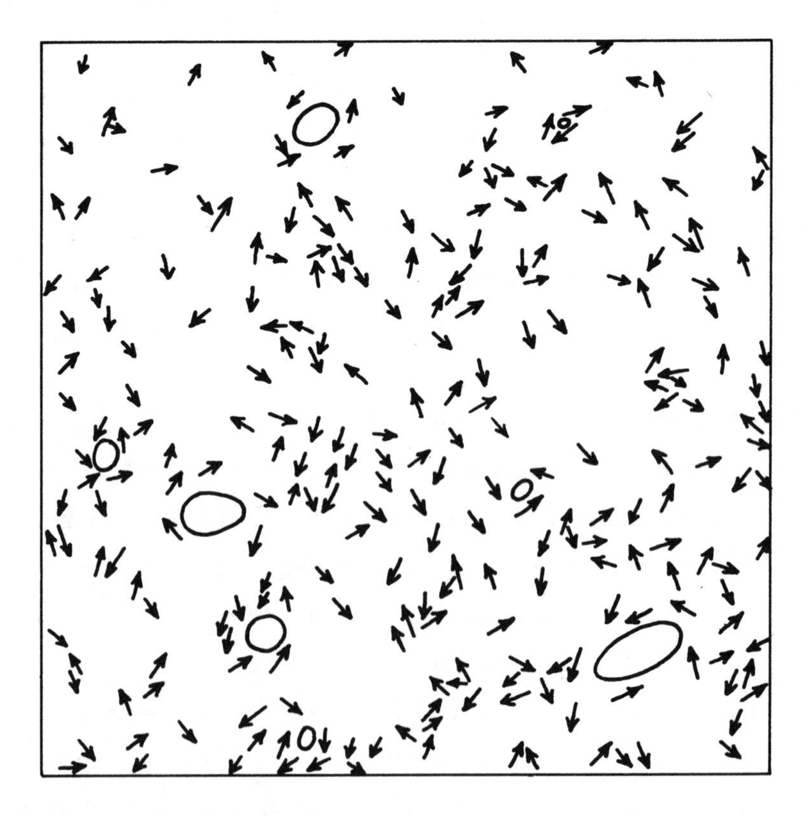

FIGURE 9.26. $T=t_{92}$

CHAPTER X - DISCRETE SPECIAL RELATIVITY

10.1 INTRODUCTION

Thus far, emphasis has been directed toward Newtonian mechanics. In this chapter, we shall show that discreteness is also fundamental in special relativistic mechanics in the sense that the classical energy equation

$$(10.1) \quad E = mc^2$$

can be established using only differences and difference quotients. The discussion will depend heavily on the development of Taylor and Wheeler, who used differences, rather than derivatives, wherever possible, to reach a reader who wished to study relativity, but whose background was minimal.

10.2 BASIC CONCEPTS

Consistently, we will measure not only length, but also time, in the same unit, meters, as follows. A meter of time, denoted by 1 meter/c, is the time it takes for light to travel one meter. Thus,

$$(10.2) \quad 1 \text{ meter}/c = (3.335640)10^{-9} \text{ sec.}$$

It will be assumed that at every point in Euclidean three-space there is a clock which is calibrated in meters and is synchronized with the clock at the origin. When one observes an event and records not only its position, but also the time on the clock at that position, one says that an observation has been made in space-time. The coordinates of an event are of the form (x,y,z,t). With regard to the observation of events in space-time, it will be assumed that the coordinate system is inertial, that is, free of gravitational acceleration, and that all laws of physics and all physical constants are the same in every inertial frame.

Though time will always be measured in meters, it is sometimes convenient to measure speed conventionally as v meters per second and at other times in light-time as β meters per meter. Thus, if t_1 and t_2 are any two time readings such that

$$t_2 - t_1 = 1 \text{ meter}/c ,$$

and if a particle in motion along an X axis is at x_1 at time t_1 and at x_2 at time t_2, then we define β and v at t_1 by the forward differences

$$(10.3) \quad \beta = \frac{x_2 - x_1}{t_2 - t_1}$$

$$(10.4) \quad v = \frac{x_2 - x_1}{(t_2 - t_1)(3.335640)10^{-9}} .$$

The units of β are then meters per meter, while the units of v are meters per second. From (10.3) and (10.4) one has

$$(9.5) \quad \beta = v/c .$$

Of course, the speed of light β^* is given by

$$\beta^* = 1 \text{ meter per meter.}$$

Note also that if a particle has a <u>constant</u> speed β, then (10.3) does yield this exact value from t_1, t_2, x_1 and x_2.

10.3 INERTIAL FRAMES

Next, consider two inertial frames moving relative to each other in such a way that their X axes are collineal. Call one the laboratory frame and call the second, which moves in a positive direction relative to the first, the rocket frame. A light flashes and is recorded in both systems. The problem is to relate the coordinates (x,y,z,t) in the lab frame to the coordinates (x',y',z',t') in the rocket frame. Under the simplifying assumptions that the flash occurs on the X axes with $y = z = y' = z' = 0$, and that the origins of the two systems are coincident at $t = 0$, then, if β_r is the constant speed of the rocket frame relative to the lab frame, and if $\beta_r < 1$, the desired relationships are a special case of the well-known Lorentz transformation and are given by

$$(10.6)\quad x = [x' + \beta_r t'][1 - \beta_r^2]^{-1/2}$$

$$(10.7)\quad t = [\beta_r x' + t'][1 - \beta_r^2]^{-1/2} .$$

10.4 PROPER TIME

With regard to the time of an event, observe that the quantity τ, given by

$$(10.8)\quad \tau = [t^2 - x^2]^{1/2} ,$$

can be rewritten by means of (10.4) and (10.5) as

$$(10.9)\qquad \tau = [(t')^2 - (x')^2]^{1/2} .$$

Since τ is the same in both coordinate frames, it is an invariant which, when $t^2 - x^2 > 0$, is defined to be the proper time of an event. In observing two events, say E_1 with $x = x_1$, $t = t_1$ and E_2 with $x = x_2$, $t = t_2$, then

$$(10.10)\qquad \Delta\tau = [(t_2 - t_1)^2 - (x_2 - x_1)^2]^{1/2}$$

is called the proper time between the two events and is also an invariant of the Lorentz transformation.

10.5 ENERGY

Finally, let us now turn to the concept of energy. Consider a particle P of mass m which, for simplicity, is in motion only on an X axis of, say, a lab frame. Its position is observed at every $\Delta t = (3.335640)10^{-9}$ seconds. Let t_1 and t_2 be the times of two consecutive observations and let x_1 and x_2 be the respective X-coordinates of P at these times. Then the particle's relativistic energy E^* at time t_1 is defined by the forward difference formula

$$(10.11)\qquad E^* = m\,\frac{t_2 - t_1}{\Delta\tau} ,$$

where the units of E^* are units of mass. To convert relativistic energy E^* to energy E in conventional units requires (Taylor and Wheeler) multiplication of E^* by c^2, so that

$$(10.12)\qquad E = E^* c^2 .$$

By means of (10.5), (10.10), and (10.11), one can then rewrite (10.12) as

$$
\begin{aligned}
E &= m \frac{t_2 - t_1}{\Delta\tau} c^2 \\
&= mc^2 / \left(\frac{\Delta\tau}{t_2 - t_1} \right) \\
&= mc^2 / \left[1 - \left(\frac{x_2 - x_1}{t_2 - t_1} \right)^2 \right]^{1/2} \\
&= mc^2 / (1 - \beta^2)^{1/2} .
\end{aligned}
$$

If $\beta < 1$, then

$$
\begin{aligned}
E &= mc^2 \left(1 + \frac{\beta^2}{2} + \frac{3}{8}\beta^4 + \cdots \right) \\
&= mc^2 + \frac{m\beta^2 c^2}{2} + \cdots \\
&= mc^2 + \frac{mv^2}{2} + \cdots .
\end{aligned}
$$

For β small, then,

$$
(10.13) \quad E \sim mc^2 + \frac{mv^2}{2} ,
$$

where $mv^2/2$ is the kinetic energy of the particle and mc^2 is called its rest energy, because, when $v = 0$,

$$
(10.14) \quad E = mc^2 .
$$

Thus, the well known formula (10.14) has followed directly from difference formulations (10.3), (10.4) and (10.11) of the basic physical concepts of velocity and energy.

It should be noted that other relativistic concepts, like momentum, can be defined, similarly, in terms of forward differences.

APPENDIX A

THE VAN DER POL OSCILLATOR

In this appendix we will consider a special nonlinear oscillator which is of broad interest in circuit theory. It is called the van der Pol oscillator and it is special because one wants to study it <u>without</u> prescribing initial conditions, contrary to the spirit of Chapter II.

A van der Pol oscillator is one whose dynamical behavior is determined by

$$\text{(A.1)} \qquad a_k = \lambda(1 - x_k^2)v_k - x_k, \qquad k = 0,1,2,\ldots,n-1,$$

where λ is a nonnegative constant and, from (1.5) and (1.6),

$$\text{(A.2)} \qquad v_{k+1} = v_k + a_k\Delta t, \qquad k = 0,1,2,\ldots,n-1$$

$$\text{(A.3)} \qquad x_{k+1} = x_k + \frac{\Delta t}{2}(v_{k+1} + v_k), \quad k = 0,1,2,\ldots,n-1.$$

The physical problem of interest with regard to (A.1) is that of finding a periodic solution. Thus, one is <u>not</u> given an initial value problem. Indeed, one can consider the problem as that of finding the initial conditions so that the solution of the resulting initial value problem is periodic, and this can be done as follows. Set

(A.4) $\quad x_0 = a, \quad v_0 = 0,$

and, for various values of a, try to find t_{2k} such that from (A.1)-(A.3)

(A.5) $\quad x_{2k} = a, \quad v_{2k} = 0 .$

If this can be done, then a periodic solution, its period t_{2k}, and the initial value a will be known. We will implement this idea next, but do only half the required work, by assuming symmetry (see Urabe) of the periodic solution about t_k.

Make initial guesses $x_0^{(n)} = n+1$, $n = 0,1,2,\ldots,10$, for the constant a. For each $x_0^{(n)}$ generate, in order, the sequence $x_{k+1}^{(n)}$, $k = 0,1,2,\ldots$, from (A.1)-(A.3). Each of these sequences will decrease initially from positive values through negative values. Terminate each iteration when

$$x_{K+1}^{(n)} \geq x_K^{(n)} ,$$

that is, when each sequence stops decreasing, and record

$$S_n = x_0^{(n)} + x_K^{(n)} .$$

The finite sequence S_n, $n = 0,1,2,\ldots,10$, will be an increasing sequence which, initially, is negative. Our aim is to find an S_n which is zero. With this in mind, let $n = \mu$ be the first value of n for which

(A.6) $\quad S_\mu \cdot S_{\mu+1} \leq 0.$

The condition (A.6) implies that S_μ is non-positive while $S_{\mu+1}$ is non-negative. Then set $a = x_0^{(\mu)}$ and $t_k = K\Delta t$.

Thus, $x_0^{(\mu)}$ is an integer which approximates a. To compute a one decimal place refinement of this approximation, begin, again, by setting $x_0^{(0)} = 0.0 + x_0^{(\mu)}$, $x_0^{(1)} = 0.1 + x_0^{(\mu)}$, $x_0^{(2)} = 0.2 + x_0^{(\mu)}, \ldots, x_0^{(10)} = 1.0 + x_0^{(\mu)}$ and recycle. Thus, if one had found $x_0^{(\mu)} = 2$, one would recycle with $x_0^{(0)} = 2.0$, $x_0^{(1)} = 2.1, \ldots, x_0^{(10)} = 3.0$. From the resulting one decimal place refinement, one can continue in the fashion indicated to a j-decimal place refinement, in which j is limited only by one's computer capability.

On the UNIVAC 1108, the following approximations for a and t_k were generated by this method with $\Delta t = 0.001$:

$$\lambda = 10 \,, \qquad a = 2.014 \,, \qquad t_k = 9.538$$

$$\lambda = 1.0 \,, \qquad a = 2.009 \,, \qquad t_k = 3.335$$

$$\lambda = 0.1 \,, \qquad a = 2.000 \,, \qquad t_k = 3.148 \,.$$

The graphs of the approximate periodic functions are shown in Figure A.1.

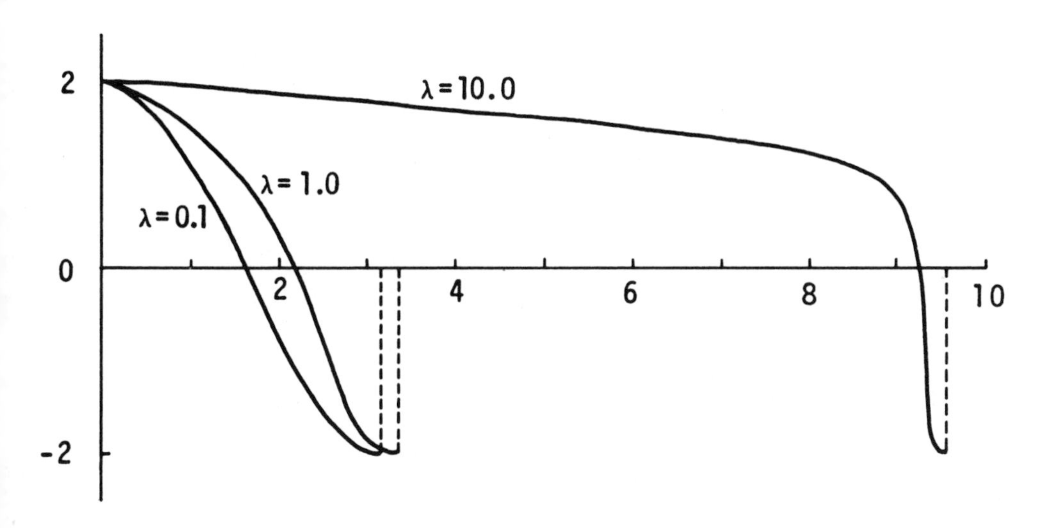

FIGURE A.1

APPENDIX B

A DISCRETE HAMILTON'S PRINCIPLE

Conservative dynamical behavior can be conceived of as a consequence of an optimization process. In such an approach to physics, Newtonian dynamical equations are deduced, rather than assumed. Since we have not stressed such an approach, and since it is of interest, we will indicate how it can be developed from the discrete point of view in this appendix. For simplicity, we will consider only the case of one dimensional motion under the force of gravity, given by (1.11).

Let the discrete Hamilton sum H_n be defined by

$$H_n = \sum_{i=0}^{n} (\tfrac{1}{2}mv_i^2 - mgx_i) , \tag{B.1}$$

and, for illustrative purposes, consider first, say, n=5. Extremization of

$$H_5 = \sum_{i=0}^{5} (\tfrac{1}{2}mv_i^2 - mgx_i) \tag{B.2}$$

implies that

$$\frac{\partial H_5}{\partial x_i} = 0 , \quad i = 1,2,3,4,5, \tag{B.3}$$

or, from (B.2), since v_i is independent of x_j for $j > i$,

$$\text{(B.4)} \quad v_5 \frac{\partial v_5}{\partial x_5} = g$$

$$\text{(B.5)} \quad v_4 \frac{\partial v_4}{\partial x_4} + v_5 \frac{\partial v_5}{\partial x_4} = g$$

$$\text{(B.6)} \quad v_3 \frac{\partial v_3}{\partial x_3} + v_4 \frac{\partial v_4}{\partial x_3} + v_5 \frac{\partial v_5}{\partial x_3} = g$$

$$\text{(B.7)} \quad v_2 \frac{\partial v_2}{\partial x_2} + v_3 \frac{\partial v_3}{\partial x_2} + v_4 \frac{\partial v_4}{\partial x_2} + v_5 \frac{\partial v_5}{\partial x_2} = g$$

$$\text{(B.8)} \quad v_1 \frac{\partial v_1}{\partial x_1} + v_2 \frac{\partial v_2}{\partial x_1} + v_3 \frac{\partial v_3}{\partial x_1} + v_4 \frac{\partial v_4}{\partial x_1} + v_5 \frac{\partial v_5}{\partial x_1} = g .$$

More precisely, (B.4)-(B.8) and Theorem 1.2 imply

$$\text{(B.9)} \quad v_5 \left(\frac{2}{\Delta t}\right) = g$$

$$\text{(B.10)} \quad v_4 \left(\frac{2}{\Delta t}\right) + v_5 \left(-\frac{4}{\Delta t}\right) = g$$

$$\text{(B.11)} \quad v_3 \left(\frac{2}{\Delta t}\right) + v_4 \left(-\frac{4}{\Delta t}\right) + v_5 \left(\frac{4}{\Delta t}\right) = g$$

$$\text{(B.12)} \quad v_2 \left(\frac{2}{\Delta t}\right) + v_3 \left(-\frac{4}{\Delta t}\right) + v_4 \left(\frac{4}{\Delta t}\right) + v_5 \left(-\frac{4}{\Delta t}\right) = g$$

$$\text{(B.13)} \quad v_1 \left(\frac{2}{\Delta t}\right) + v_2 \left(-\frac{4}{\Delta t}\right) + v_3 \left(\frac{4}{\Delta t}\right) + v_4 \left(-\frac{4}{\Delta t}\right) + v_5 \left(\frac{4}{\Delta t}\right) = g$$

the unique solution of which is

$$\text{(B.14)} \quad v_5 = g\Delta t/2,\ v_4 = 3g\Delta t/2,\ v_3 = 5g\Delta t/2,\ v_2 = 7g\Delta t/2,\ v_1 = 9g\Delta t/2.$$

Thus, (B.14) and (1.6) yield

$$a_1 = -g, \quad a_2 = -g, \quad a_3 = -g, \quad a_4 = -g,$$

or,

$$(B.15) \quad ma_1 = -mg, \; ma_2 = -mg, \; ma_3 = -mg, \; ma_4 = -mg.$$

Finally, from (1.11) and (B.15), one has the discrete Newtonian equation

$$(B.16) \quad F_i = ma_i, \quad i = 1,2,3,4.$$

In the fashion indicated above, it follows readily with respect to (B.1) that

$$F_i = ma_i, \quad i = 1,2,\ldots,n-1,$$

that is, that Newton's equation is valid at each time step between, but not including, t_0 and t_n, as a consequence of extremization of H_n.

APPENDIX C

CONSERVATION OF ANGULAR MOMENTUM IN THREE DIMENSIONS

Because of our emphasis on computability, conservation of energy has been proved wherever possible, since it implies a certain type of numerical stability. Occasionally, we have proved also the conservation of linear momentum. In this appendix we will show how to establish in discrete form the third basic conservation principle, that of angular momentum, and because of its importance in three dimensional problems, we will do it for this case by vector methods.

If, at time t_j, particle P_i of mass m_i is located at $\vec{x}_{i,j}$ and has velocity $\vec{v}_{i,j}$, then its angular momentum $\vec{L}_{i,j}$ is defined by

$$\text{(C.1)} \qquad \vec{L}_{i,j} = m_i(\vec{x}_{i,j} \times \vec{v}_{i,j}) .$$

Hence, from the three dimensional forms of (4.1)-(4.3),

$$\vec{L}_{i,j+1}-\vec{L}_{i,j} = m_i(\vec{x}_{i,j+1}\times\vec{v}_{i,j+1}) - m_i(\vec{x}_{i,j}\times\vec{v}_{i,j})$$

$$= m_i\left[(\vec{x}_{i,j+1}-\vec{x}_{i,j}) \times \left(\frac{\vec{v}_{i,j+1}+\vec{v}_{i,j}}{2}\right)\right.$$

$$\left. + \left(\frac{\vec{x}_{i,j+1}+\vec{x}_{i,j}}{2}\right) \times (\vec{v}_{i,j+1}-\vec{v}_{i,j})\right]$$

$$= m_i\left[(\vec{x}_{i,j+1}-\vec{x}_{i,j}) \times \left(\frac{\vec{x}_{i,j+1}-\vec{x}_{i,j}}{\Delta t}\right)\right.$$

$$\left. + \left(\frac{\vec{x}_{i,j+1}+\vec{x}_{i,j}}{2}\right) \times (\vec{a}_{i,j}\Delta t)\right]$$

$$= \Delta t \left(\frac{\vec{x}_{i,j+1}+\vec{x}_{i,j}}{2}\right) \times \vec{F}_{i,j} \ .$$

Thus,

$$\text{(C.2)} \qquad \vec{L}_{i,j+1} - \vec{L}_{i,j} = (\Delta t)\vec{T}_{i,j}$$

where the vector

$$\vec{T}_{i,j} = \frac{\vec{x}_{i,j+1}+\vec{x}_{i,j}}{2} \times \vec{\mathbf{F}}_{i,j}$$

is called the torque of P_i at t_j.

If for system P_i, $i = 1,2,\ldots,n$, one defines angular momentum $\vec{L}_j$ at t_j by

$$\vec{L}_j = \sum_{i=1}^{n} \vec{L}_{i,j}$$

and torque $\vec{T}_j$ by

$$\vec{T}_j = \sum_{i=1}^{n} \vec{T}_{i,j} \ ,$$

then, from (C.2),

(C.3) $\vec{L}_{j+1} - \vec{L}_j = (\Delta t)\vec{T}_j$.

Hence, for any system in which $\vec{T}_j = \vec{0}$ for $j = 0,1,2,\ldots,$ one has

$$\vec{L}_{j+1} = \vec{L}_j , \quad j = 0,1,2,\ldots,$$

or, equivalently,

(C.4) $\vec{L}_j = \vec{L}_0 , \quad j = 1,2,3,\ldots,$

which is called the law of conservation of angular momentum.

RESEARCH PROBLEMS

1. Establish sufficient conditions under which a solution of algebraic system (3.5) exists and is unique.

2. Develop sufficient conditions for the stability of the discrete string of Example 1, Section 3.4.

3. Construct an example of a discrete vibrating string which consists of 10^6 particles. Describe the motion which results after release from various initial positions of tension.

4. Develop a discrete string model in which the particles are arranged in a more complex configuration than that shown in Figure 3.1. For example, allow the particles to be arranged in clusters.

5. Prove the existence or nonexistence of an explicit, energy conserving formulation of discrete gravitation.

6. Establish general conditions under which a solution of algebraic system (4.18)-(4.21) exists, is unique, and can be determined by Newton's method.

7. Are there any mathematical relationships between stability and energy conservation for the discrete model of gravitation of Chapter IV?

8. Establish whether or not the error bounds claimed by G. M. Clemence, in his calculation of the perihelion motion of Mercury, are valid.

9. Determine the effects of varying solar and planetary shapes and masses on the perihelion motion of Mercury.

10. Develop a discrete model of the von Karman vortex street.

11. Develop a discrete model of convective heat transfer.

12. Develop a discrete model of friction.

13. Develop a discrete model of elastic buckling.

14. Develop another (see, e.g., Saltzer) discrete model of potential theory.

15. Develop a discrete model of boundary layer formation.

16. Develop a discrete, unifying theory of boundary layers and shock waves.

17. Develop a discrete calculus of variations (see, e.g., Cadzow) which nowhere uses derivatives.

18. Develop discrete Lagrangians and Hamiltonians.

19. From the classical point of view, the discrete formulations of Chapter IX are all of second order. Are there any higher order formulations which are energy conserving with respect to gravity?

20. Establish whether or not the implicit method of Section 9.2 is stable for all Δt and all α when applied to (2.1).

21. Develop, completely, a discrete Special Relativity.

22. Develop a discrete General Relativity.

23. Construct a realistic example of von Neuman's recommendation for the discrete modeling of fluids, in which relatively few particles are used and the dynamical parameters are adjusted to maintain hydrodynamic balance.

REFERENCES AND SOURCES FOR FURTHER READING

B. J. Alder
"Studies in molecular dynamics III: A mixture of hard spheres", Jour. Chem. Phys., 40, 1964, pp. 2724-2730.

R. L. Berger and N. Davids
"General computer method of analysis of conduction and diffusion in biological systems with distributive sources", Rev. Sci. Instr., 36, 1965, pp. 88-93.

A. S. Besicovitch
"On the definition and value of the area of a surface", Quart. Jour. Math., 16, 1945, pp. 86-102.

J. A. Cadzow
"Discrete calculus of variations", Int. Jour. Control, 11, 1970, pp. 393-407.

G. F. Carrier
"On the nonlinear vibration problem of an elastic string", Quart. Appl. Math., 3, 1945, pp. 157-165.

G. M. Clemence
"The motion of Mercury 1765-1937", Astron. Papers Amer. Ephem. and Naut. Almanac, XI, U.S. Govt. Printing Off., Wash., D. C., 1943.

R. Courant and K. O. Friedrichs
Supersonic Flow and Shock Waves, Interscience, N. Y., 1948.

C. W. Cryer
"Stability analysis in discrete mechanics", TR 67, Dept. Comp. Sci., U. Wis., Madison, 1969.

N. Davids and N. E. Kesti
"Stress-wave effects in the design of long bars and stepped shafts", Int. Jour. Mech. Sci., 7, 1965, pp. 759-769.

N. Davids, P. K. Mehta, and O. T. Johnson
"Spherical elastoplastic waves in materials", in Behavior of Materials under Dynamic Loading, ASME, N. Y., 1959, pp. 125-137.

C. R. Deeter
"Discrete generalized functions", Jour. Math. Anal. Appl., 39, 1972, pp. 375-396.

C. R. Deeter and G. Springer
"Discrete harmonic kernels", Jour. Math. Mech., 14, 1965, pp. 413-438.

R. J. Duffin
"Basic properties of discrete analytic functions", Duke Math. Jour., 23, 1956, pp. 335-363.

E. Fermi, J. R. Pasta, and S. Ulam
"Studies of nonlinear problems I", TR 1940, Los Alamos Sci. Labs., Los Alamos, N. M., 1955.

R. D. Feynman, R. B. Leighton, and M. Sands
The Feynman Lectures on Physics, Addison-Wesley, Reading, Mass., 1963.

D. Greenspan

1. "Discrete mechanics", TR 49, Dept. Comp. Sci., U. Wis., Madison, 1968.
2. "Discrete, nonlinear string vibrations", The Comp. Jour., 2, 1970, pp. 195-201.
3. "Discrete liquid flow", TR 14, Computing Center, U. Wis., Madison, 1970.
4. "Discrete shock waves", TR 84, Dept. Comp. Sci., U. Wis., Madison, 1970.
5. "Numerical approximation of periodic solutions of van der Pol's equation", Jour. Math. Anal. Appl., 39, 1972, pp. 574-579.

6. Introduction to Numerical Analysis and Applications, Markham, Chicago, 1971.
7. "Numerical studies of the 3-body problem", SIAM Jour. Appl. Math., 20, 1971, pp. 67-78.
8. "Computer simulation of transverse string vibrations", BIT, 11, 1971, pp. 399-408.
9. "A new explicit discrete mechanics with applications", Jour. Franklin Inst., 294, 1972, pp. 231-240.
10. "New forms of discrete mechanics", Kybernetes, 1, 1972, pp. 87-101.
11. "The harmonic oscillator", TR 130, Part I, Dept. Comp. Sci., U. Wis., Madison, 1971.
12. "A discrete theory of Newtonian gravitation", TR 130, Part II, Dept. Comp. Sci., U. Wis., Madison, 1971.
13. "Symmetry in discrete mechanics", to appear in Found. Phys.
14. "A finite difference proof that $E=mc^2$", Amer. Math. Mo., 80, 1973, pp. 289-292.
15. "A discrete approach to fluid dynamics", Proc. IFIPS 71, North-Holland, Amsterdam, 1972.
16. "Progression and regression in the perihelion motion of the planet Mercury and proposed computer tests related to general relativity", TR 26, Comp. Ctr., U. Wis., Madison, 1971.
17. "Discrete Newtonian gravitation and the n-body problem", Util. Math., 2, 1972, pp. 105-126.
18. "A numerical study of the mixing of fluids with an example of discrete turbulence", TR 153, Dept. Comp. Sci., U. Wis., 1972.
19. "Discrete bars, conductive heat transfer, and elasticity", TR 164, Dept. Comp. Sci., U. Wis., 1972, to appear in Computers and Structures.
20. "Discrete solitary waves", TR 167, Dept. Comp. Sci., U. Wis., 1972.
21. "An algebraic, energy conserving formulation of classical molecular and Newtonian n-body interaction", to appear in Bulletin, Amer. Math. Soc., March, 1973.
22. "An arithmetic, particle theory of fluid dynamics", TR 171, Dept. Comp. Sci., U. Wis., 1973.

P. R. Halmos
Measure Theory, van Nostrand, N. Y., 1950.

W. Hayes and R. F. Probstein
Hypersonic Flow Theory, Academic Press, N. Y., 1959.

J. O. Hirschfelder, C. F. Curtis, and R. B. Bird
Molecular Theory of Gases and Liquids, Wiley, N. Y., 1954.

P. Henrici
Error Propagation for Difference Methods, Wiley, N. Y., 1963.

P. R. Jonas and J. T. Bartlett
"The numerical simulation of particle motion in a homogeneous field of turbulence", J. Comp. Phys., 9, 1972, pp. 290-302.

S. T. Jones
"Fortran program for discrete conductive heat transfer" Appendix, TR 164, Dept. Comp. Sci., Univ. Wis., 1972.

T. von Karman
Aerodynamics, McGraw-Hill, N. Y., 1963.

J. G. Kemeny and J. L. Snell
Mathematical Models in the Social Sciences, Ginn and Co., N. Y., 1962.

W. S. Krogdahl
"Numerical solutions of the van der Pol equation", ZAMP, 11, 1960, pp. 59-63.

R. E. Langer
"Fourier series", Amer. Math. Mo., 57, II, 1947.

A. K. MacPherson
"The formulation of shock waves in a dense gas using a molecular dynamics type technique", Jour. Fluid Mech., 45, 1971, pp. 601-621.

P. K. Mehta
"Cylindrical and spherical elastoplastic stress waves by a unified direct analysis method", AIAA Journal, 5, 1967, pp. 2242-2248.

R. H. Miller, K. H. Prendergast, and W. J. Quirk
"Numerical experiments on spiral structure", COO-614-72, Inst. Comp. Res., Univ. Chicago, no date.

R. von Mises
Mathematical Theory of Compressible Fluid Flow, Academic Press, N. Y., 1958.

R. E. Moore
Interval Analysis, Prentice-Hall, Englewood Cliffs, N. J., 1966.

J. von Neumann
"Proposal and analysis of a new numerical method for the treatment of hydrodynamical shock problems", in The Collected Works of John von Neumann, vol. 6, Pergamon, N. Y., 1963.

Ju. G. Ostapov
"Stability of the motions of discrete dynamical systems", Mathematical Physics, No. 6, 1969, pp. 149-157 (Russian) Naukova Dumka, Kiev, 1969.

J. R. Pasta and S. Ulam
"Heuristic numerical work in some problems of hydrodynamics", MTAC, 13, 1959, pp. 1-12.

Yu. P. Popov and A. A. Samarskii
"Completely conservative difference schemes for the equation of gas dyanmics in Euler's variables", USSR Comp. Math. and Math. Phys., 10, 1970, pp. 265-273.

R. W. Preisendorfer
Radiative Transfer on Discrete Spaces, Permagon, N. Y., 1965.

D. Raftopoulos and N. Davids
Elastoplastic impact on rigid targets", AIAA Jour., 5, 1967, pp. 2254-2260.

L. M. Rauch and W. C. Riddell
"Iterative solutions of the analytical n-body problem", SIAM Jour. Appl. Math., 8, 1960, pp. 568-581.

P. Richman
"ε-Calculus", TR 105, Dept. Comp. Sci., Stanford Univ., Palo Alto, Calif., 1968.

R. D. Richtmyer and K. W. Morton
Difference Methods for Initial-Value Problems, 2nd ed., Wiley, N. Y., 1967.

B. Russell
Principles of Mathematics, 2nd ed., W. W. Norton, N. Y., 1938.

P. G. Saffman
"Lectures on homogeneous turbulence", in Topics in Nonlinear Physics, N. J. Zabusky, Ed., Springer, N.Y., 1968.

H. Salacker and E. P. Wigner
"Quantum limitations of the measurement of space-time distances", Phys. Rev., 109, 1958, pp. 571-577.

C. Saltzer
"Discrete potential and boundary value problems", Duke Math. Jour., 31, 1964, pp. 299-320.

H. Schlichting
Boundary Layer Theory, 4th Ed., McGraw-Hill, N. Y., 1960.

A. B. Schubert
"FORTRAN program for the mixing of fluids", Appendix, TR 153, Dept. Comp. Sci., Univ. Wis., 1972.

A. B. Schubert and D. Greenspan
"Numerical studies of discrete vibrating strings", TR 158, Dept. Comp. Sci., Univ. Wis., 1972.

D. Schultz
"Programming string vibrations", Appendix, TR 57, Dept. Comp. Sci., Univ. Wis., Madison, 1969.

K. Symon
Mechanics, Addison-Wesley, Reading, Mass., 1960.

K. Symon, D. Marshall, and K. W. Li
"Bit-pushing and distribution-pushing techniques for the solution of the Vlasov equation", PLP 383, Plasma Studies, Univ. Wis., 1970.

J. L. Synge and B. A. Griffith
Principles of Mechanics, McGraw-Hill, N. Y., 1942.

E. F. Taylor and J. A. Wheeler
Spacetime Physics, Freeman, San Francisco, 1966.

M. Urabe
"Galerkin's procedure for non-linear periodic systems and its extension to multi-point boundary value problems for general nonlinear systems", in Numerical Solutions of Nonlinear Differential Equations, Wiley, N. Y., 1966, pp. 296-328.

A. N. Whitehead
Process and Reality, Macmillan, N. Y., 1929.

A. N. Whitehead and B. Russell
Principia Mathematica, 2nd ed., vol. 1, Cambridge Univ. Press, Cambridge, 1925.

N. J. Zabusky
"Elastic solutions for the vibrations of a nonlinear continuous model string", Jour. Mathematical Phys., 3, 1962, pp. 1028-1039.

SUBJECT INDEX